The California Electricity Crisis: Causes and Policy Options

• • •

Christopher Weare

2003

PUBLIC POLICY INSTITUTE OF CALIFORNIA

Library of Congress Cataloging-in-Publication Data
Weare, Christopher.
  The California electricity crisis: causes and policy options /
  Christopher Weare.
      p. cm.
  Includes bibliographical references.
  ISBN: 1-58213-064-7
  1. Electric utilities—Government policy—California. 2. Electric
industries—California. 3. Energy policy—California.
I. Title.

  HD9685.U53 C294 2002
  333.793'2'09794—dc21   2002151734

Foreword

Understanding the interplay of events behind California's recent energy crisis is a formidable challenge. Even more formidable is imagining the policy changes that can rebuild the state's energy markets. Christopher Weare's report, *The California Electricity Crisis: Causes and Policy Options*, addresses both of these challenges. It serves as a useful guide to a complex chain of events as well as a helpful description of options that state officials will weigh as they design and implement the next set of policy solutions.

Policymakers and general audiences alike can draw several lessons from Weare's analysis. The first is that energy policy is forged out of a complex blend of technical, economic, political, and historical realities. Energy provision, pricing, and distribution are determined by what the engineers know is possible, what the regulators think should be done, and what the politicians want to see. This complexity makes it difficult to implement sweeping changes without generating unintended consequences. As Weare points out, such consequences impose costs of their own, not all of which are well understood when the initial reforms are proposed and implemented.

Related to this first lesson is the possibility that frustrated observers will propose simplistic solutions to complex problems. Some may even try to implement their reforms through the initiative process. If residents wish to avoid price swings in their electricity bills, a proposition to this effect could gain widespread support. Such solutions, however, could make efficient and low-cost energy even more difficult to provide. One gathers from Weare's analysis that accommodating the intricacies of this market and crafting effective solutions will be a difficult task no matter who controls the policy levers in Sacramento.

A second lesson is that the federal government (and especially the Federal Energy Regulatory Commission) will provide the framework for any subsequent energy policy. For over a decade now, politicians have successfully pushed for the devolution of political power to state and local governments. However, if California chooses to reconstitute competitive energy markets, it will have to accept federal review. That process will be no less complicated than balancing the local, state, and federal interests that have accompanied efforts to create efficient water markets in California.

Finally, the electricity crisis has reminded us that Californians—like most Americans—do not like unpleasant surprises. Blackouts, price volatility, excess profits, poor service, and vague promises have combined to reinforce the public's view that Sacramento cannot be trusted. To regain the public's trust, decisionmakers must explain their objectives and then craft a sensible, sustainable policy in a timely fashion. Otherwise, simplistic and possibly draconian solutions may begin to gather support. We trust that this report and its recommendations will help policymakers in their deliberations and planning.

David W. Lyon
President and CEO
Public Policy Institute of California

Summary

With the passage of AB 1890 in 1996, California led the nation in efforts to deregulate the electricity sector. The act was hailed as a historic reform that would reward consumers with lower prices, reinvigorate California's then-flagging economy, and provide a model for other states. Six years later, the reforms lay in ruins, overwhelmed by electricity shortages and skyrocketing prices for wholesale power. The utilities were pushed to the brink of insolvency and are only slowly regaining their financial footing. The state became the buyer of last resort, draining the general fund and committing itself to spending $42 billion more on long-term power deals that stretch over the next ten years. The main institutions of the competitive market established by AB 1890, the Power Exchange and retail choice in particular, have been dismantled.

The debate over the exact causes of the crisis continues. Many wish to distill the genesis of the crisis to simple themes. Some, most notably major political actors in California, lay principal blame on market manipulation by the merchant generators. Others, including the Federal Energy Regulatory Commission and energy firms, point to flaws in the state's restructuring plan and a fundamental supply and demand imbalance. Any search for simple answers, however, risks misperceiving the intricacies of the systemic failure of California's electricity sector. A satisfactory explanation for the severity of the crisis and its consequences cannot be composed based on any single factor. Rather, a number of factors must be considered. These include:

- A shortage of generating capacity,
- Bottlenecks in related markets,
- Wholesale generator market power,
- Regulatory missteps, and
- Faulty market design.

No single factor can fully account for the crisis. The fault cannot be pinned entirely on the shortage in generating capacity. The worst of the crisis occurred during the winter of 2000–2001, when demand was low and plenty of capacity should have been available. Similarly, market manipulation by generators does not tell the whole story. There is evidence of the exercise of market power, but increased input costs and demand also pushed market prices higher. Although the division of regulatory authority between California and the federal government led to catastrophic policy paralysis in response to the crisis, it cannot be blamed for the run-up in wholesale rates that instigated the crisis. Finally, flaws in the restructuring of the electricity sector did exacerbate the crisis, but the market had been working reasonably well for the first two years of its operations.

Because California's experience was unique and because a number of factors were simultaneously at play, it is not possible to disentangle fully how each distinctly contributed to the blackouts, major financial crisis, and the systemic breakdown of market institutions. Some important conclusions can, nevertheless, be offered.

First, California's electricity sector was rocked by a number of events unrelated to restructuring: the rise in national natural gas prices, higher costs for pollution permits, and a drought in the Northwest which reduced available imports of electricity. Even if the electricity sector had remained regulated, prices would have increased, and some blackouts would have possibly occurred between May 2000 and June 2001. Second, although regulators have yet to uncover a smoking gun clearly establishing that merchant generators strategically manipulated wholesale market prices, market and regulatory conditions created an environment ripe for the exercise of market power.[1] The shortages in generating capacity played a critical role, increasing the bargaining strength of merchant generators and signaling the enormous profits that could be gained through supply shortages. At the same time, the excessive reliance

---

[1] Recently, regulators have uncovered evidence of market manipulation strategies employed by Enron and other electricity trading firms. These strategies, however, targeted small ancillary markets, such as those that manage congestion on transmission lines. They did not uncover any evidence of manipulation of the main market for wholesale power.

on the spot market increased the opportunities and incentives for generators to increase their prices well above the costs of generating power. Third, California relied far too much on the spot market for wholesale power instead of securing power through more stable long-term contracts. This choice exposed the utilities to exceptional risks, producing a full-blown financial fiasco. Finally, the division in regulatory authority between state and federal regulators impeded policymakers from developing a rapid, coordinated, and effective response before major damage was inflicted on the electricity sector, the California economy, and all Californians.

Because the crisis has left California's energy sector in such disarray, policymakers face the daunting task of reconstructing the market and regulatory institutions of the electricity sector almost entirely from scratch. Decisions over the long-run institutional structure of California's electricity sector are complicated by the complexity of the issues that the crisis unearthed and the wide range of options being debated. Serious proposals representing almost the entire spectrum of economic philosophies are receiving significant attention. These include calls for increased public ownership of the electricity sector, a return to the system of regulated, vertically integrated utilities, and recommendations for further deregulation. We examine the costs and benefits of these major options, focusing on six primary goals for the electricity sector:

- Low prices,
- Stable bills for customers,
- Efficient use of resources by producers and consumers,
- A reliable supply of electricity,
- Administrative feasibility, and
- Protection of the environment.

Overall, policymakers face a choice between the greater stability, reliability, and administrative feasibility provided by public ownership or regulated regimes versus the prospects for greater efficiency gains through competitive markets. In terms of environmental protections, no regime clearly dominates the others, mainly because environmental results depend on complex interactions between each regime and existing environmental regulations.

Eventually, movement to reinstate elements of the competitive regime, in particular competitive wholesale generation, is almost inevitable. The federal government continues to push for greater wholesale competition through the creation of regional trading organizations. In addition, technological advances create ever-smaller plants that can generate electricity at competitive costs, facilitating entry by new firms and enabling large customers to self-generate. Efforts to bottle up these sources of power through public ownership or regulation become increasingly difficult and inefficient. In the short run, policymakers may choose to restrain the development of competitive generation markets if they wish to promote a more stable electricity sector and are wary about ceding control to the Federal Energy Regulatory Commission for mitigating the market power of competitive generators. Nevertheless, they should exercise caution in making short-run choices that erect barriers against loosening these constraints on competition in the future.

On the retail side of the market, the tradeoffs between regulated and competitive structures depend on consumers. Potential efficiency gains from competition are derived by changing consumer behavior, making them more aware of the real costs of electricity and allowing them to change their consumption accordingly. These gains can come about, however, only if consumers are exposed to price volatility and are willing and able to manage that volatility. If consumers wish to be shielded from such volatility and wish to remain passive consumers of energy, the benefits of a competitive regime are reduced. Concerns over the ability of consumers to manage electricity price volatility suggest that hybrid models that introduce retail competition in stages, first to larger customers and only later to smaller customers, offer important advantages.

The report also offers three recommendations for policy changes that can improve the performance of the electricity sector under any particular regulatory and market structure. The first is to strengthen and institutionalize demand-management programs. Electricity sector restructuring ignored and often undermined demand-side management.

Regulators failed to promote retail competition. Funding for conservation programs was reduced, and consumers were shielded from price fluctuations. As policymakers continue to seek ways to balance the supplies and demands within California's electricity sector, demand management cannot be left out of the equation. These programs can lower energy costs, improve efficiency, and enhance system reliability. In addition, promoting demand management can make individuals and firms more intelligent consumers of electricity, facilitating the introduction of retail competition and enabling them to benefit from competitive offerings.

The second recommendation is to develop a capacity for more comprehensive planning and oversight of California's energy infrastructure. Inadequate transmission capacity, overreliance on natural gas plants, bottlenecks in natural gas pipelines, and inadequate natural gas storage all contributed to the state's troubles. An overarching review of these interlocking infrastructure components is necessary to ensure that private investments are adequate and to identify areas in which public investment or coordination is required.

The third recommendation is to reassess and reorganize the complex set of administrative structures that currently exist. Electricity sector restructuring followed by crisis has led to an ad hoc and confusing mix of state agencies and departments. This fractured and overlapping set of agencies leads to inefficiencies, conflicts, and policy confusion. It must be redesigned for effective policy development and implementation and to provide a more certain environment for producers and consumers.

California policymakers need to take away a number of hard-earned lessons from the crisis. The complexity of electricity markets cannot be underestimated, and seemingly inconsequential details of market design can have significant and unexpected consequences. Specifically, heavy reliance on spot markets is extraordinarily risky. Policymakers must also appreciate the extent to which the state's control over the electricity sector has been circumscribed by the split of regulatory authority between the state and federal governments. Finally, if market-based reforms are to be successful, firms and consumers must become more responsive to market incentives and risks. During the restructuring of the electricity sector, however, utilities and consumers continued to operate as if the stable and secure rules of regulation still held, leaving California woefully unprepared for the price spikes in 2000.

At this juncture, policymakers must focus on forging a consensus on the future direction of California's electricity sector. Continued ambiguity and conflict lead to market uncertainty, stifle investment in critical infrastructure, and risk repeating errors that precipitated the crisis. Agreement on the broad outlines of a regulatory and market structure, even without the details specified, would do much to improve the investment environment and enable California to move forward.

Contents

Figures

Tables

Acronyms

AB	Assembly bill
CEC	California Energy Commission
CERA	Cambridge Energy Research Associates
CPA	California Consumer Power and Conservation Financing Authority, also known as the California Power Authority
CPUC	California Public Utilities Commission
CTC	Competitive transition charge
DWP	Department of Water and Power
DWR	Department of Water Resources
EOB	Electricity Oversight Board
ESP	Electricity service provider
FERC	Federal Energy Regulatory Commission
GWh	Gigawatt hour
ISO	Independent System Operator
kWh	Kilowatt hour
MMBtu	Million British thermal units
MMcf	Million cubic feet
MW	Megawatt
MWh	Megawatt hour
NOx	Nitrogen oxide
PG&E	Pacific Gas & Electric

PX	Power Exchange
QF	Qualifying facilities
RECLAIM	Regional Clean Air Incentives Market
RTP	Real-time pricing or real-time prices
SCAQMD	South Coast Air Quality Management District
SCE	Southern California Edison
SDG&E	San Diego Gas & Electric
TOU	Time of use

1. Introduction

In 1996, California passed AB 1890, a bill calling for the radical restructuring of the state's electricity sector. Competitive markets for wholesale power were inaugurated in April 1998, and in those early years, the markets appeared to function relatively well. As predicted, the wholesale price of electricity declined and average rates fluctuated moderately between $20 and $50 per megawatt hour (MWh) (see Figure 1.1). Customers benefited from a 10 percent rate reduction and were protected by a temporary rate freeze. The utilities benefited at the same time, as they were able to pay off the costs of transitioning to a competitive environment.

In the late spring of 2000, however, the electricity sector began to malfunction severely. In June, average prices suddenly rose precipitously, breaking the $100 per MWh mark. They remained at extraordinarily high rates through the spring of 2001 before they moderated rapidly and

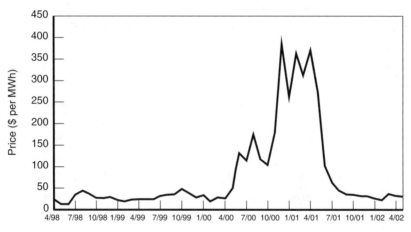

SOURCE: Joskow and Kahn (2001b).

Figure 1.1—Average Wholesale Electricity Prices in California, 1998–2002

unexpectedly in June 2001 (see Figure 1.1). Although total energy costs for wholesale power were $7.4 billion in 1999, they were about $27 billion per year from 2000 through 2001, burdening California consumers and businesses with almost $40 billion in added costs.

The lights flickered throughout the crisis. On June 14, 2000, rolling blackouts in San Francisco caused by a Bay Area heat wave signaled the beginning of rough times. In 2000, electricity was turned off to customers with special interruptible contracts on 13 other days. During 2001, "load shedding" occurred on 31 days. On nine of these days customers experienced involuntary rolling blackouts for a total of 42 hours of outages. During these nine outages, California experienced an average shortfall of 600 MW of electricity, enough energy to power over 450,000 households. On the worst day, January 18, the equivalent of almost one million households lost electricity. The costs of these blackouts are difficult to enumerate, but they are undoubtedly significant.

The soaring prices on the wholesale market wreaked financial havoc on the electricity sector. The customers of San Diego Gas & Electric (SDG&E) felt the brunt of the cost increases immediately. The retail rate freeze imposed on the utilities had been lifted for SDG&E in July 1999. Thus, SDG&E customers were paying electricity rates based on wholesale prices and saw their bills double and triple during the summer of 2000. Customers of Pacific Gas & Electric (PG&E) and Southern California Edison (SCE), in contrast, were shielded from these increases by the retail rate freeze. These two utilities, however, were caught in a financial vise, forced to buy expensive power on the wholesale market and sell it cheaply to retail customers. Soon, SDG&E joined them in this predicament when the legislature passed AB 265, which reimposed a rate freeze for SDG&E customers retroactively.[1] The three major utilities racked up debt at a rapid pace. In January, as their credit worthiness evaporated, the state was forced to become the purchaser of last resort.

---

[1]AB 265 included provisions to enable SDG&E to recoup the uncompensated costs of buying wholesale power. Thus, it was not placed in the same financial peril as were PG&E and SCE.

A long list of debts is still being sorted out. Pacific Gas & Electric declared bankruptcy and is arranging in bankruptcy court how to pay creditors about $13 billion. Southern California Edison accepted a deal with the California Public Utilities Commission (CPUC) in which it will pay off $5 billion to $6 billion in debt with a combination of ratepayer contributions, cash on hand, and decreased dividends. The state spent $8.7 billion on wholesale power in the first half of 2001 and projected that it would spend $17.2 billion by the end of the year. $7 billion for these purchases came from the general fund, and the state is still struggling to float a $12 billion bond to repay the fund. In addition, during the height of the crisis the state began signing long-term contracts for power to secure a source of supply, and it is now committed to purchase $42 billion worth of electricity over the next ten years.

Beyond this financial turmoil, the crisis caused by the surge in wholesale prices devastated the institutional structures governing the California electricity sector. The private utilities are no longer the main purchasers of power. Instead, the state is more tightly entwined in the electricity market than it has ever been before. The Power Exchange (PX), the central market for trading wholesale power, went bankrupt and closed operations. The Independent System Operator (ISO), designed to manage the electricity grid, has become politicized and is under fire. The state has curtailed retail choice, putting competition on hold, and regulatory authority is now more fragmented, leading to overlaps and conflict. The destruction wrought by the financial crisis and system failure has been so complete that California must re-create the regulatory and market institutions of its electricity sector almost from scratch.

To gain some perspective on the damage inflicted on the California economy, one can compare it with other significant economic failures. This crisis has cost $40 billion in added energy costs over the last two years. Increased costs will continue as long as the prices in the long-term contracts signed by the state exceed wholesale rates. On top of these costs, one must add the costs of blackouts and reductions in economic growth caused by the crisis.[2] Thus, conservatively, the total costs can be

---

[2]The national recession has complicated estimating the macroeconomic effects of the crisis, but in June UCLA projected that the crisis would slow the California economy

placed around $40 billion to $45 billion or around 3.5 percent of the yearly total economic output of California. Before this crisis, the preeminent example of failure of an electricity system was a default by the Washington Public Power Supply System. It overinvested in nuclear plants and defaulted on its bonds. This default cost the state about $800 million or 1.5 percent of its total economic output. The Savings and Loan debacle was considered a staggering deregulatory failure, but its total costs of about $100 billion amounted to only one-half of 1 percent of the total U.S. economy.

Repairing this damage poses a daunting task to California policymakers. Much of the debate and legislative action has focused on the financial dimensions of the crisis. In contrast, the manner in which the state is going to extricate itself from its role as the power purchaser of last resort, reorganize the electricity sector, and regulate it remains imprecise. This report seeks to focus attention on these important institutional questions.

After a brief overview of the regulatory reforms that led to this crisis, this report examines the root causes of the crisis. It finds that blame cannot be easily leveled at any single actor. A combination of unforeseen events, poor decisions, opportunistic behavior, and fragmented regulatory authority all conspired to aggravate the magnitude of the crisis.

Based on this analysis of the root causes of the crisis, Chapter 4 of the report examines a number of frameworks that may guide the reorganization of the electricity sector: increased public ownership, return to a regulated environment, continuing with competitive markets, and hybrids of these options. It concludes that some form of competition should be reinstated, at least for certain industry segments and customer classes. In the short run, however, policymakers may choose to curtail the role of competition for the sake of stability and

---

in 2002 by between 0.7 and 1.5 percent and would increase unemployment by 1.1 percent. See Cambridge Energy Research Associates (2001b).

administrative ease and to provide a smoother transition path back to a competitive environment. Chapter 5 then discusses specific policy options that are appropriate no matter which reform path is chosen.

2. Regulatory Context

Before restructuring, the California electricity sector was dominated by three investor-owned utilities: Pacific Gas & Electric, Southern California Edison, and San Diego Gas & Electric. Together they accounted for 77 percent of customers and 75 percent of all state power sales in 1996. The rest of the industry was composed of four cooperatives and 34 publicly owned utilities, run mainly by municipalities or irrigation districts. The Los Angeles Department of Water and Power (DWP) and the Sacramento Municipal Utility District were the two largest, accounting for approximately 15 percent of California sales.

California differs from the rest of the United States in some important dimensions. California is a light consumer of electricity. Consumption is only 6,400 kilowatt hours (kWh) per resident per year compared to the national average of 11,900 per capita. Similarly, as a percentage of its economic output, California is efficient, requiring only 0.22 kWh for each dollar of state output compared to 0.40 kWh for the country as a whole. California also differs in the mix of generating plants on which it depends. As seen in Table 2.1, California has virtually no coal generation but this is the primary source of electricity for the nation as a whole. The state is richer in inexpensive hydro generation than the rest of the nation but it is much less reliant on it than Oregon and Washington. California has also invested heavily in renewable sources. Most striking is California's heavy reliance on natural gas. Including plants fired by a combination of petroleum and natural gas, over 50 percent of California's electricity comes from this source, compared to only 18 percent nationally.

In the early 1990s, the CPUC began to explore the possibility of restructuring the state's electricity sector to open it up to competitive forces. In February 1993, it issued a report, commonly known as the yellow book, which promoted regulatory reform. As the first serious

Table 2.1

Generating Capacity by Primary Energy Source, 1999
(in percent)

Energy Source	California	United States	Oregon	Washington	Nevada	Arizona
Hydro	27	16[a]	82	84	15	19
Nuclear	8	14	0	4		25
Natural gas	36	18	11	5	35	20
Petroleum/gas combo	16	0	0	1	3	0
Coal	1	44	5	5	44	35
Petroleum	2	9	0	0	1	2
Other[b]	11	0	2	1	3	0

SOURCE: http://www.eia.doe.gov/emeu/states/_states.html.

[a]U.S. hydro includes renewables.

[b]Geothermal, wind, and solar.

effort by state-level policymakers to consider a major restructuring of the entire electricity sector, the report marked a major milestone in the deregulatory movement in the United States. Electricity markets and regulation had, nevertheless, been undergoing slow but steady liberalization for decades.

In response to the energy shortages of the 1970s, Congress had passed the Public Utility Regulatory Policies Act (PURPA) in 1978. It allowed nonutilities to build qualifying facilities (QFs) that generated electricity with cogeneration technologies or renewable resources and required that utilities purchase this energy. Further actions by Congress and the Federal Energy Regulatory Commission (FERC) continued to loosen the domination of traditional utilities. Congress passed the Energy Policy Act of 1992, allowing independent firms, called exempt wholesale generators (EWGs), to operate generation facilities. FERC issued a number of decisions that granted independent power producers access to the electricity transmission grid that was largely owned by the major, private utilities. By allowing generators to sell electricity to faraway customers, these actions facilitated the development of a fringe market for wholesale power. As seen in Figure 2.1, the percentage of California power generated by nonutilities increased from less than 5

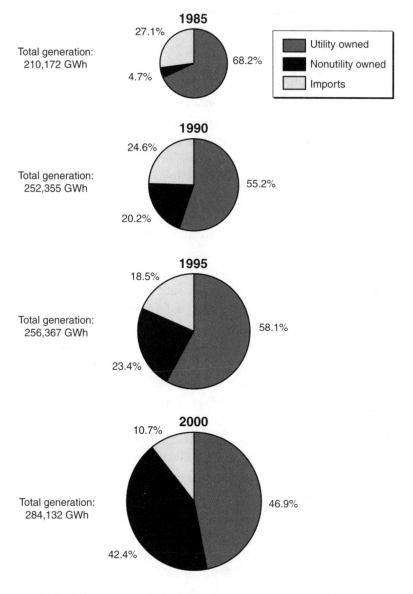

1985

Total generation:
210,172 GWh

27.1%

4.7%

68.2%

Utility owned

Nonutility owned

Imports

1990

Total generation:
252,355 GWh

24.6%

20.2%

55.2%

1995

Total generation:
256,367 GWh

18.5%

23.4%

58.1%

2000

Total generation:
284,132 GWh

10.7%

42.4%

46.9%

SOURCE: California Energy Commission, http://www.energy.ca.gov/electricity.

Figure 2.1—Total California Electricity Generation

percent in 1985 to over 23 percent in 1995. Nevertheless, traditional electricity monopolies, regulated by state bodies, continued to control the lion's share of generating plant.

In the early 1990s, interest in restructuring the California electricity sector was spurred by the high cost of electricity. In 1995, because of expensive investments in nuclear power and high-priced contracts for QF power, California consumers paid the highest rates in the western continental United States. The average rate of about 9.9 cents per kWh was more than twice as much as the rates in Oregon and Washington, 60 percent more than that in Nevada, and 30 percent more than that in Arizona. Businesses and industry began to see that new generating facilities could supply electricity at lower prices than the utilities were charging, and they wished to take advantage of these lower costs. Moreover, an increasing number of commentators began to argue that the traditional system of supplying electricity through a vertically integrated, regulated monopoly was a source of these problems. They argued that the regulatory regime provided insufficient incentives to control costs, led to excess generating capacity, and resulted in unwise investments in nuclear power. The United States was benefiting from the success of recent experiments in deregulating the airline, trucking, natural gas, and long-distance telephone industries. The electricity industry was a natural extension of the deregulatory model.

The CPUC held protracted hearings for several years and developed a detailed proposal for deregulation. At that juncture, the legislature became involved, drafting AB 1890, a blueprint for electricity sector reform. The legislation incorporated the central elements of the CPUC plan that restructured the three major, vertically integrated investor-owned utilities. As shown in Figure 2.2, before AB 1890, they owned generation plants, transmission lines, and distribution facilities and marketed all their own electricity. On the fringe, a number of qualifying facilities and exempt wholesale generators produced power and sold it to the three major utilities or to municipal utilities.

AB 1890 sought to break up the utilities and create competitive markets in both the generation and the retail marketing of electricity. As shown in Figure 2.2, competitive wholesale generators were invited into the market. To jump-start competition, the California utilities were

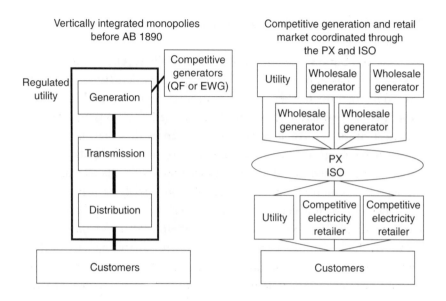

Figure 2.2—AB 1890 Restructured Electricity Sector

given incentives to divest generation facilities. They eventually sold all their fossil fuel plants, almost 19,000 MW of generating capacity, predominantly to five merchant generation firms: Southern Energy, Duke Energy Corporation, Houston Industries, AES Corporation, and NRG Energy.[1] These plants represented 43 percent of the utilities' generating capacity before AB 1890 and were 35 percent of the state's total generating capacity. The utilities retained their nuclear and hydro facilities and their contracts with QFs, but their share of total generation dropped below 50 percent as seen in Figure 2.1.

The utilities handed over the control of their transmission grids to a newly created, independent, nonprofit organization, the California ISO. The ISO was to manage the dispatch of electricity through the grid in a nondiscriminatory manner providing equal access to all generators and power purchasers. In addition, it would maintain system reliability by balancing the demand and supply of electricity in real time. Competitively generated power would be sold through the newly created

---

[1] PG&E also sold 1,353 MW of geothermal generating capacity to Calpine Corporation.

Power Exchange (PX), which ran auctions for power in day-ahead and hour-ahead markets.

The distribution networks of the old vertically regulated utilities remained regulated, but new electricity service providers (ESPs) would be allowed to enter the market, sign up customers, and provide them with power that they would purchase on the open market. It was expected that most customers would eventually be served by these new ESPs. These arrangements roughly followed the design of the deregulated system in the United Kingdom that inspired the California reform, although it was more ambitious and relied more heavily on markets.

Parallel to the restructuring of the electricity market, regulatory oversight was also fundamentally changed. Before AB 1890, the California Public Utilities Commission was the primary regulator of the state's vertically integrated utilities. It accounted for the utilities' costs of generation, transmission, and distribution and set retail rates that enabled the utilities to recoup those costs with an allowed return on invested capital. The FERC regulated wholesale power sales and purchases by the utilities, but its role was peripheral to state regulatory commissions. As AB 1890 elevated the competitive wholesale market to a central role in the restructured electricity sector, FERC assumed a dominant regulatory position, overseeing the operation of the PX and ISO and regulating rates for wholesale power and transmission. California's role, in contrast, was diminished, limited to the regulation of retail rates and distribution.

The legislature added a number of pertinent features to AB 1890 designed to satisfy major stakeholders. A roadblock to regulatory reform was the existence of stranded costs—reductions in the value of utility assets caused by the transition from a regulated to a competitive environment. Wholesale prices were expected to decrease after deregulation, leaving the utilities unable to recover unamortized investments they had made as regulated monopolies. These included generating facilities, mostly nuclear plants, and high-priced, long-term contracts with QFs. Utilities resisted deregulation as long as it would force them to write off these costs. To placate the utilities, AB 1890 allowed them to recoup these stranded costs through a state bond issue and a competitive transition charge (CTC). In the interest of consumers, AB 1890 cut retail rates by 10 percent (about the same amount as the

CTC) and froze rates until utilities had completed paying off their stranded costs. In addition, supporters of the environment and conservation received some subsidies. Overall, the legislation promised benefits for industry, small customers, utilities, and the environment. It passed both houses of the legislature unanimously and was signed by Governor Pete Wilson on September 23, 1996.

3. Root Causes of the Electricity Crisis

The causes and consequences of the crisis are multiple, complex, and intertwined, but there is wide agreement concerning the broad causal factors of the crisis. Almost unanimously, analysts cite five significant factors:

1. A shortage of generating capacity,
2. Bottlenecks in related markets,
3. Wholesale generator market power,
4. Regulatory missteps, and
5. Faulty market design.

There remains significant debate over the relative importance of each of these factors. Some, most notably major political actors in California, wish to lay principal blame on market manipulation by the merchant generators. Others, including the Federal Energy Regulatory Commission and energy firms, have pointed mainly to flaws in the state's restructuring plan. Any search for simple answers, however, risks misperceiving the intricacies of the systemic failure of California's electricity sector. A satisfactory explanation for the severity of the crisis and its consequences cannot be composed based on any single factor. All of these factors contributed and reinforced one another to create a unique and explosive combination.

A Shortage of Generating Capacity

The tight supply of electricity generating capacity beginning in the summer of 2000 appears to be the primary cause of the California electricity crisis. The evidence indicates that tight supply was a necessary antecedent to the crisis. During the early years of the wholesale market, electricity supply was ample, and the market worked reasonably well.[1] Evidence from other markets also indicates that markets are most competitive when there is ample supply to meet demand, but as the supply of electricity tightens markets become less competitive (Bushnell and Saravia, 2002).

The statistics clearly demonstrate an increasingly tight electricity market.[2] Total consumption in California steadily increased by about 1.5 percent per year between 1990 and 2000 (see Figure 3.1). In addition, there was a surge in demand of about 4 percent per year between 1998 and 2000, driven by the then-booming economy. It is important to note, however, that growth in demand during the 1990s was actually lower than the rate of growth during the 1980s, even when considering the effects of increased population and economic activity (Brown and Koomey, 2002).

The growth rates in neighboring states were significantly higher. Nevada's electricity demand grew at a 6.2 percent yearly rate between 1988 and 1998, and Arizona's demand grew 3.7 percent per year. Higher demand in neighboring states is significant because California had historically relied on imports from other states for about 20 percent

---

[1]A 1999 conference at the University of Southern California evaluated restructuring and gave it positive marks (USC School of Policy Planning and Development, 1999). Other early analyses of the California experiment note problems with market design (Joskow, 2000; Hogan, 2001b). However, these authors indicate that these issues were being addressed, and they find no evidence of an impending crisis.

[2]It does require some care, however, to interpret and compare the various statistics on energy consumption and generation. First, generation capacity (measured in MW) and total generation (measured in MWh) numbers are often used interchangeably. Although they are related, the main focus should be on generation capacity available for peak demands. Because the load profile for California appears to be flattening (e.g., the average electricity use of all hours in a day is closer to the peak demand), the effects of increased total use on peak demand have been mitigated. Also, researchers employ statistics from a number of agencies (e.g., the California Energy Commission (CEC), ISOs, proprietary data, and the U.S. Energy Information Administration). They also employ statistics for different geographical service areas (e.g., the western region, California, and California served by utilities). There are disparities between these statistics, and a detailed understanding of how they are collected is required before they can be compared correctly.

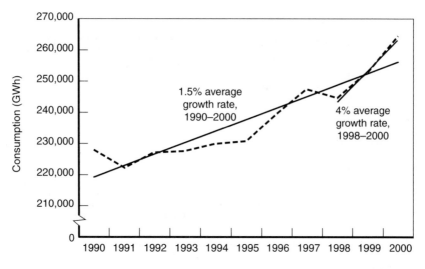

Figure 3.1—Total Electricity Consumption in California, 1990–2000

of its electricity needs. Thus, a major source of supply was being eaten up by growth outside California.

California Energy Commission data show that peak demand—the total amount of electricity consumed during the one-hour period each year that experiences the highest demand—grew more slowly between 1990 and 2000, with an annual growth rate of 1.0 percent (see Figure 3.2). The peak demand increased more slowly for reasons that remain obscure. The rise of the high-tech sector may have increased the demand for the continuous operation of computers and networks, thereby increasing energy use relatively more during nonpeak, night hours.

Despite this growth in demand, capacity remained stagnant (see Figure 3.2). Consequently, reserves—the amount of generating capacity available above current demand—fell from over 12 percent to less than 5 percent of capacity (see Figure 3.3). Reserves are essential for maintaining reliable delivery of electricity by enabling the system operator to react to surges in demand or failures in either generating stations or transmission links. The California ISO, for example, calls an emergency Stage 1 alert and requests users to reduce power demand

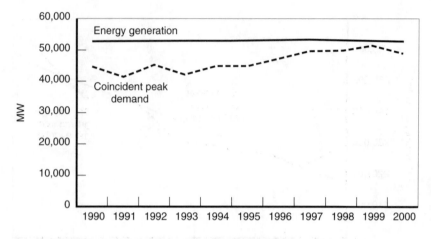

SOURCE: California Energy Commission, http://www.energy.ca.gov/electricity.

Figure 3.2—Peak Demand and Generation Capacity in California,
1990–2000

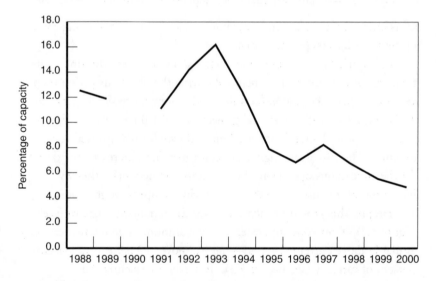

SOURCE: California Energy Commission, http://www.energy.ca.gov/electricity.

Figure 3.3—Total Reserves for Yearly Peak Demand in California,
1988–2001

whenever reserves fall below 7 percent. If reserves fall below 1.5 percent, the ISO begins to implement rolling blackouts. Traditional industry practice has been to maintain a reserve margin of 15 percent or more, although California had been able to maintain lower in-state reserves because of the availability of imported power.

Although the shortage in generating capacity is not disputed, the ultimate factors that led to the shortage are more controversial. Finger-pointing flourished as commentators blamed deregulation, inaction by the governor, generating firms, and the design of market institutions. Each of these explanations contains an element of truth, but responsibility for the shortage cannot easily be placed on the shoulders of any single actor or institution. The shortage was largely a historical accident, characterized by a unique confluence of factors. A number of unforeseen events combined to lay the foundations for a supply crunch, and regulatory and market failures exacerbated the shortage. Each of these factors is examined in turn.

Unforeseen Events

In a well-functioning electricity system, either regulated or market based, it is important to maintain a balance between available generating capacity and peak electricity demand. Excess capacity is unwanted because it increases the average cost of electricity generation, and insufficient generating capacity leads to the risk of blackouts. This balance, however, can be easily upset in the short term because demand for electricity can change much more quickly than the time it takes to design, gain approval for, and construct new generating plants.

Shortages have been rare in the United States because an emphasis on reliability has meant that added capacity was built well in advance of need. Nevertheless, short-term problems have occurred. In 1948, after a spurt of rapid post–World War II growth and an extended drought, Northern California experienced a series of blackouts before rains refilled reservoirs (Ross, 1974). In the late 1990s, a confluence of unexpected events combined to produce a similar short-lived imbalance in generating capacity. Market players were caught by surprise by the surge in demand in California and throughout the West in the late 1990s. Generating firms were in the process of planning and building additional capacity.

Between 1997 and 2000, they filed applications to build nearly 15,000 MW of generating capacity (see Figure 3.4). Generator firms, however, appear to have been planning that market demand would outstrip available supply only in 2001 and later. None of these plants were scheduled for completion in 2000, and less than 2,000 MW were scheduled to be available by the summer of 2001. Because of faulty market expectations, these new plants arrived later than needed, leading to interim shortages.

The California Energy Commission contests this story, pointing to forecasts dating as far back as 1988 that correctly predicted demand for 2000 and 2001 (California Energy Commission, 2001). The implication is that industry insiders should not have been surprised by the demand for electricity and should have been investing to meet expected demand. The CEC forecasts, however, did not consider how unexpectedly strong

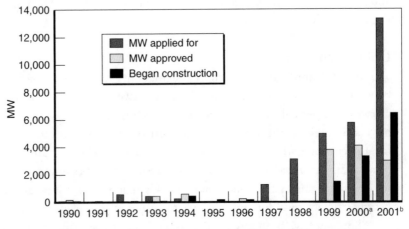

SOURCE: California Energy Commission, http://www.energy.ca.gov/sitingcases/projects_since_1976.html.

[a]In 2000, 16 plant applications were recorded for 5,740 MW. Of those, eight applications for 1,184 MW were withdrawn. One plant for 99 MW was withdrawn after approval.

[b]In 2001, 45 plant applications were recorded for 13,309 MW. Of those, eight applications for 1,509.6 MW were withdrawn. Two plants for 242.4 MW began construction but were later withdrawn (they are not included in the number of MW that began construction).

Figure 3.4—Generation Applications and Approvals, 1990–2001

demand growth outside California would reduce the availability of imported power.

California and the West were also hit by unfavorable weather conditions. The winters of both 2000 and 2001 were relatively dry in the West, and in particular in the Northwest (see Figure 3.5). Consequently, the amount of hydro generation available was severely reduced. Estimates indicate that in the summer of 2000, there were 8,000–12,000 fewer megawatts of hydro power available for import into California, representing up to 20 percent of California's summer demand (California State Auditor, 2001a, p.59). In addition, these conditions were combined with an unusually hot summer in 2000 that drove up electricity demand throughout the western United States (see Figure 3.6). As shown in Figure 3.7, these conditions combined to reduce electricity imports to California to their lowest levels in ten years.

From the summer of 2000 through the winter of 2001, supply shortfalls were exacerbated by unscheduled outages of generating facilities (see Figure 3.8). These high levels of outages were to some degree coincidental. They were due in part to poor coordination of standard

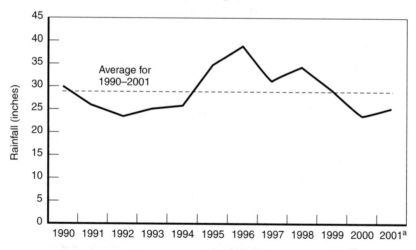

SOURCE: Western Regional Climate Center, http://www.wrcc.dri.edu.
[a]Does not include December.

Figure 3.5—Average Yearly Pacific Northwest Rainfall, 1990–2001

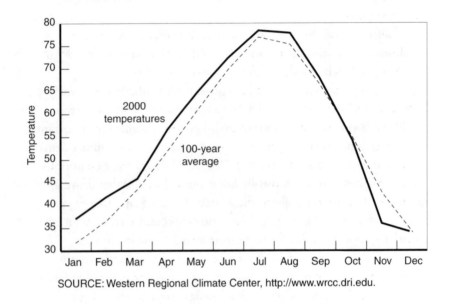

SOURCE: Western Regional Climate Center, http://www.wrcc.dri.edu.

Figure 3.6—Monthly Temperatures in the Western United States,
100-Year Average and 2000

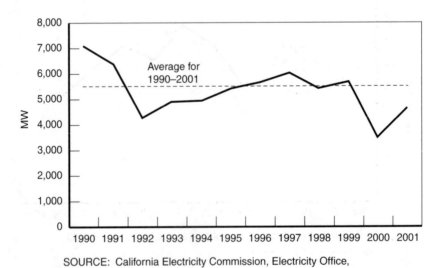

SOURCE: California Electricity Commission, Electricity Office,
http://www.energy.ca.gov/electricity/electricity_generation.html.

Figure 3.7—Total Imported Electricity for California, 1990–2001

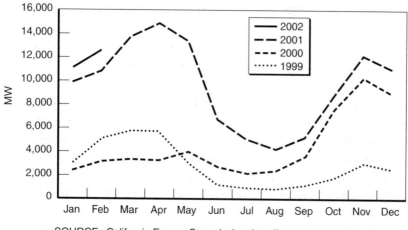

SOURCE: California Energy Commission, http://www.energy.ca.gov/
electricity/1999-2001_monthly_off_line.html.

**Figure 3.8—Average Daily Forced or Scheduled Megawatts Off-Line by
Month, 1990–2002**

maintenance. For example, in the fall of 2000, 5,000 MW of power was
taken off-line when nuclear plants scheduled maintenance at the same
time. Plant operators have also pointed to the age of generation plants
and to the fact that they were suffering from deferred maintenance
because of heavy use in the early months of 2000. As will be discussed
below, others disagree, contending that producers were strategically
withholding their plants from the market in an effort to exercise market
power and increase the prices they received.

Regulatory Failures

The regulatory environment following the passage of AB 1890 also
contributed to the tightening supply situation. During the
implementation of the market restructuring, regulatory uncertainty was
pervasive as market rules were amended frequently. In addition,
Proposition 9 was placed on the ballot in November 1998. It aimed to
restrict the payments the utilities could receive for their stranded costs,
thereby clouding the future of the California deregulation. It was
eventually defeated by more than a 2-to-1 margin but, nevertheless, it

diverted attention away from plant construction at an important juncture.

This uncertainty was not unique to California. Throughout the United States, federal and state regulators were moving away from cost-based regulation of vertically integrated monopolies. As part of this transition, utilities abandoned their traditional role of constructing capacity. Since the mid-1980s, virtually all capacity additions were constructed by unregulated wholesale generators, but the rules under which these wholesale generators would operate remained incomplete, leading, at least in part, to a more general decline in investment throughout the United States (Joskow, 2000, p. 153; Bushnell and Saravia, 2002).

Several reports have also criticized the environmental review and siting process as too long and too expensive, hindering investment (Bay Area Economic Forum, 2001a; Cambridge Energy Research Associates, 2001a). For example, the average length of the permitting process in California is 14 months compared to only seven in Texas (Bay Area Economic Forum, 2001a; Smith, 2001). The California State Auditor found that for projects proposed since the beginning of 1997, the CEC missed its own one-year deadline in 11 out of 15 cases (California State Auditor, 2001b). Public opposition to projects did contribute to delays but it was not the only impediment to approval. The applicants themselves caused significant delays as they amended applications and submitted documents late. Other government agencies, such as air quality districts, also contributed to the length of the process, forcing the CEC to wait for necessary inputs.

The record of the last few years does suggest that regulatory uncertainty and a slow review process delayed the construction of new capacity. The California State Auditor identified over 1,000 MW of capacity that could have been available during the critical months in the late spring and early summer of 2001 if the CEC had completed siting review in a timely manner. Nevertheless, these environmental regulations do not appear to have been a critical impediment to investment. The average length of siting reviews since the passage of AB 1890 has not been appreciably longer than reviews in the 1970s and 1980s. A review took on average one and one-half months longer, and much of this increase can be accounted for by the fact that more recent projects tended to involve fossil fuels rather than solar and geothermal energy and were submitted by firms with less experience with the siting process (California State Auditor, 2001b). In any case, as is shown by Figure 3.4, even during the early, more uncertain years of the California electricity market, investors still submitted applications to increase California electricity generating capacity by over 20 percent.

Other regulatory decisions not related to environmental review may have had a greater effect in reducing the number of projects undertaken. For reasons discussed in greater detail below, the California Public Utilities Commission issued a number of decisions that restricted the ability of the three main utilities to enter into long-term, bilateral contracts with electricity generators. Firms seeking to construct electricity generators commonly use such contracts to assure banks and potential investors that there is a market for additional electric power and that the proposed plant will be profitable. These restrictions on the utilities removed the main potential purchasers of long-run contracts from the market, increasing the risk of construction, and limiting the amount of financing available (Bay Area Economic Forum, 2001a; Cambridge Energy Research Associates, 2001a). Although these restrictions likely reduced the level of planned investment, the number of forgone projects is not known.

At the same time that regulators may have impeded private sector decisionmaking, they were relinquishing their role in the energy planning process. Before deregulation, the California Energy Commission produced comprehensive biannual evaluations of the state of the California electricity sector, but its role in resource planning was significantly diminished with the advent of deregulation. At the same time, conservation efforts diminished. Before deregulation, California had a number of innovative programs that provided utilities with incentives to invest in conservation in lieu of capacity expansion. In the early 1990s, these mandated comprehensive reviews of energy alternatives identified numerous cost-effective investments in energy-saving technologies and conservation programs (Mowris, 1998). Utilities invested as much as $400 million a year to promote these investments and programs. In the move to a competitive environment, these programs lost their constituency and momentum, largely because regulators lost the policy levers with which they provided incentives. AB 1890 did provide for continued support of such investments, but only at a much reduced level—about $220 million per year (Harvey et al., 2001).

Market Failures

Causes for underinvestment in generating capacity can also be found in the structure and operation of electricity markets.[3] Generators may have been reluctant to invest in capacity because the market failed to send strong signals that additional investments were required. The California market relied completely on spot market prices for wholesale power to signal that future investments in capacity would be profitable. In theory, when spot market prices increased or were projected to increase, generators would come forward with new investments. For the first two years after the deregulation, however, a glut of electricity drove the wholesale price to low levels. Until May of 2000, the average wholesale price of a megawatt hovered around $30 and never rose above $50. If generators focused myopically on current prices, they had little incentive to undertake new projects until supplies tightened and the spot market increased.

Because of the nature of the electricity markets, however, spot market prices are quite volatile when supplies become scarce (Borenstein, 2001). Electricity cannot be stored and there are strict constraints on the amount of electricity that can be generated and delivered at a particular time. In addition, in the short run, consumers have limited options for reducing their electricity consumption. Consequently, when an electricity market nears its maximum capacity, extreme price spikes can occur before generators bring on additional supplies, consumers reduce demand, and the market is brought back into equilibrium.

These characteristics of electricity markets can lead to boom and bust cycles in investment. Low prices impede investment until the market

---

[3]A highly contentious debate centers on the question of market power—whether suppliers withheld capacity from the market in an effort to increase prices and profits. We discuss this issue below.

tightens, leading to marked increases in prices. A rush of investments induced by high prices then frequently leads to over capacity and collapsing prices. This cycle has been evident since the crisis abated in the summer of 2001. After wholesale prices plummeted, investments in generating capacity became less attractive, and since the collapse of Enron Corporation, investors have shied away from committing their capital to large electricity-related investments. As a consequence, plans to build several new power plants have been cancelled or postponed, raising the prospect that California may experience new shortages as soon as 2004 (Tucker, 2002). Other markets, such as commercial real estate and microchips, which also involve large, capital-intensive investment with long planning horizons, have experienced such swings in capacity.

In sum, much of the blame for the dramatic increases in wholesale electricity prices experienced during the summer of 2000 can be related to inadequate supply. Although merchant generating firms had been applying for permits and had begun construction of additional capacity, these investments were slowed by the low levels of wholesale prices before the summer of 2000, regulatory uncertainty, a sluggish environmental review process, and financing impediments. It is not possible to determine the number of megawatts of additional capacity that could have been made available earlier absent these impediments, but a reasonable lower-bound estimate would be that the 1,415 MW of capacity completed in 2001 for which application had been filed in 1997 and 1998 could have been available sooner.

Unforeseen events, in contrast, had clearer and quite significant effects on reducing available supply. Reduced imports, resulting from weather conditions and increased demand in neighboring states, had the largest effect, reducing available capacity by the equivalent of up to 12,000 MW. Poor coordination of plant outages decreased available supply by about another 1,000 MW during the summer of 2000. These supply reductions constituted a significant portion of the approximately 50,000 MW peak summer demand. These events alone turned a situation of adequate supply into a shortage. It is possible that more favorable weather conditions, a downturn in the state's economy, or better operation of existing capacity could have saved California from its crisis during the summer of 2000. Then the thousands of megawatts of capacity scheduled to come on-line in 2001 and 2002 may have maintained an adequate balance between supply and demand.

This shortage, however, fails to explain the entire crisis. In a well-functioning, competitive market with fixed capacity, one would expect price spikes when demand reaches system capacity, typically on a hot summer afternoon, but prices should drop during winter months when demand is slack. As Figure 1.1 shows, however, wholesale prices rose to even higher levels in late 2000 just as demand was decreasing. The tight supplies created the conditions for these spikes, but other factors came into play.

Bottlenecks in Related Markets

Scarce electricity generating capacity was not the only shortage leading to unprecedented prices for wholesale electricity. Constraints on related infrastructure and markets, including natural gas pipelines, the market for pollution permits, and the electricity transmission system, also drove prices up.

The price of natural gas plays a key role in electricity markets. Because it is relatively clean and inexpensive, it is the fuel of choice for new generation capacity. Virtually all new generation plants burn natural gas, and the proportion of California's in-state production generated with gas plants increased from 23 percent in 1983 to 38 percent in 2000. More important, the costs of gas peaker plants, designed to run only at times of very high demand, often set the market price for electricity because of their relatively high variable costs.

From January through October 2000, an unusually cold winter on the East Coast led to a doubling of the price for natural gas at Henry Hub, Louisiana, considered an indicator of the national price of gas (see Figure 3.9). Because the variable cost of electricity is almost completely determined by gas prices, this increase also doubled generation costs. Even if California's electricity sector had remained completely regulated, the prices of electricity would have increased to reflect these increases in input costs.

Then in November 2000, the California natural gas market was hit with unprecedented volatility. Under normal circumstances, the price differential between the national and the California markets is quite

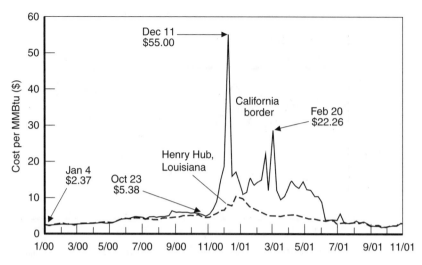

SOURCE: Enerfax Daily, http://www.sagewave.com/D2/Include/
Frame.asp?key=86575.

Figure 3.9—Natural Gas Prices, 2000–2001

small, representing the transportation costs from gas fields in the center of the United States to the California border. Yet in November, California prices rose far above the national price, and from November 2000 to June 2001, California natural gas customers were paying two to three times the national price.

As with the run-up in electricity prices, the causes of high gas prices are complex—a combination of unexpected events, demand growth, and market manipulation. An explosion in an El Paso gas pipeline on August 19, 2000, temporarily closed that source of supply, cutting off almost 15 percent of California's pipeline capacity for about ten days and reducing flows for over a year. Normally during the summer and early fall, gas is stored underground in California in preparation for winter heating demand, but this pipeline interruption reduced storage. Storage was further limited by unusually high summer demand for natural gas because of heavy use of gas-fired electricity generators to replace imports of electricity. California entered the winter of 2000–2001 with a shortfall of gas storage equal to two weeks of demand, judging by the average of the three previous years. Then, as shown in Figure 3.10,

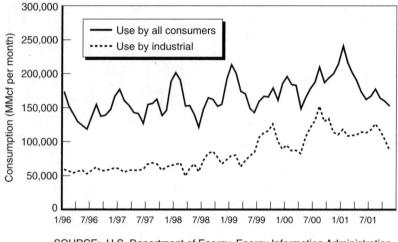

SOURCE: U.S. Department of Energy, Energy Information Administration.

Figure 3.10—California Natural Gas Consumption, January 1996 to October 2001

demand for gas reached all-time highs driven by industrial users including electricity generators.[4]

Under these circumstances—reduced supply and increased demand—an increase in gas prices is to be expected even in fully competitive markets, but these circumstances do not fully account for the much higher rates that California paid in comparison to the rates in the rest of the country. Before the summer of 2000, California had excess gas pipeline capacity. If this capacity was available, higher prices in California should have drawn in additional supplies from the Midwest until the premiums paid by California consumers were reduced.

The California Public Utilities Commission among others has charged that the El Paso Corporation strategically manipulated its control of the pipeline to drive up prices. FERC has already found that El Paso entered into an unlawful contract with one of its unregulated affiliated companies to control a significant portion of its pipeline capacity. It is further alleged that El Paso took a number of actions to

---

[4]A portion of the increase in industrial use, roughly 30,000 MMcf, is due to the transfer of gas-fired generating assets from the utilities to merchant generators.

reduce the amount of gas transported through the pipeline in an effort to constrain supply, increase prices, and boost the profits earned by its affiliate. This affiliate contract expired at the end of May 2001 and, as seen in Figure 3.9, the premiums paid by California evaporated soon thereafter. FERC has yet to promulgate a final ruling on these allegations, but an August 2002 FERC staff report found preliminary indications that manipulation of gas prices at the California border may have occurred.

During the summer of 2000, the prices for pollution permits also rose precipitously. Under an innovative pollution control system called RECLAIM (Regional Clean Air Incentives Market), industrial plants in the South Coast Air Quality Management District (SCAQMD) that emit nitrous oxides (NOx) must purchase permits for each ton of emissions. The average price of these permits rose from about $1–$2 per pound to $5 per pound in August of 2000, and some transactions were being consummated at prices exceeding $30 per pound (see Figure 3.11). This increase was caused by a combination of lower availability and higher demand. SCAQMD lowers the number of permits available every

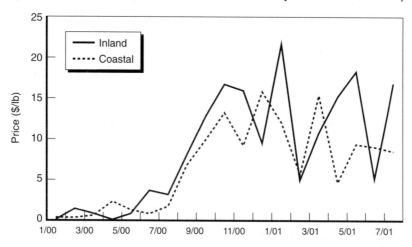

SOURCE: South Coast Air Quality Management District, Regional Clean Air Incentives Market, Trading Bulletin Board.

Figure 3.11—Average Monthly NOx Permit Price for Coastal and Inland Areas, January 2000 to July 2001

year, and 2000 was the first year in which the total number of permits constrained emissions significantly. At the same time, added generation from high-polluting plants that are not normally run for long periods increased demand for permits. Increases in NOx and natural gas prices had a dramatic effect on the marginal cost of electricity generation. Joskow and Kahn (2001b) estimate that these increased input costs raised the marginal cost of a megawatt of electricity from under $50 in May 2000 to over $100 in September 2000. Cambridge Energy Research Associates (CERA) estimate the change in marginal costs between December 1999 and December 2000 and find a much larger increase— from the $19–$35 range to the $83–$521 range.

Limitations of the electrical transmission system also contributed to the crisis. Transmission lines have fixed capacities, at times limiting the ability of energy to be shipped from generating plants to final consumers. One main constraint is Path 15, which connects Northern and Southern California. Congestion on Path 15 can prevent inexpensive power in one part of the state from being shipped to areas being served by more expensive power. This limitation has cost California electricity consumers hundreds of millions of dollars in increased energy costs and has caused rolling blackouts. San Francisco, in particular, has limited connections to the statewide grid because it is situated at the end of a peninsula. Consequently, it is vulnerable to rolling blackouts even when energy is plentiful elsewhere in the state.

These related events demonstrate that California's energy crisis extends beyond the market for wholesale power. California is facing a number of intertwined infrastructure issues. The performance of the electricity sector is dependent on, among other things, gas pipeline and storage systems, the electricity transmission system, and environmental goals. Constraints in any of these systems can dramatically and unexpectedly undermine the performance of interdependent systems, impeding the delivery of energy to California consumers.

Wholesale Generator Market Power

The shifts in market fundamentals described above—increased demand and input costs combined with decreased supply—certainly contributed to the increase in wholesale prices, but these factors cannot account for the full magnitude of the price spikes that plagued California. The highest profile and most controversial issue of the crisis has been the allegations of market manipulation and price gouging by generating firms. Early in the crisis, California appealed for relief to the Federal Energy Regulatory Commission, which regulates wholesale prices, and is pursuing $8.9 billion in refunds from energy marketers through a proceeding. More recently, California moved to nullify the long-term contracts entered into with generators during the height of the crisis, arguing that it signed the contracts under duress because of market manipulation, and during the summer of 2002, it renegotiated several of these contracts with the active encouragement of FERC's chief administrative law judge. The attorney general has filed numerous suits alleging monopolistic activities, and the California Senate has been holding months of special hearings investigating market manipulation.

The theory of market power is straightforward. The electricity generation market is dominated by a small number of producers, five national unregulated generation firms and a few public providers such as the Los Angeles Department of Water and Power and the Bonneville Power Authority. In such an oligopolistic market, generators may raise prices above the competitive level (e.g., the industry's cost for the last megawatt of generation produced) by strategically withholding some capacity from the market. When the added revenues from selling electricity at a higher price exceed the lost revenues from not selling all the electricity they could produce, their profits increase. Competition is the main barrier to such exercise of market power. When a generator with unused capacity can profitably undersell the market price, it has incentives to provide more electricity to the market, undercutting other firms' ability to raise prices.

The critical role of competition broke down during the California crisis for a number of reasons. Because of the small number of firms involved in California's energy market, it may have been possible for generators to collude, either tacitly or explicitly, to withhold capacity. Even if each generator would have individually benefited by providing additional supplies to the market, they were jointly better off by agreeing not to compete keenly.

Moreover, electricity markets are particularly prone to the exercise of market power. New supplies cannot be made available quickly because electricity cannot be easily stored, its transportation is limited by the constraints of the transmission grid, and the construction of new capacity entails long lead times. Consequently, generators do not have to worry that new supplies will flow into the market, undercutting their high bids. In addition, demand for electricity is not easily curtailed in the short run because electricity is essential to modern life and because most decisions that determine electricity usage involve the purchase of long-lived appliances (e.g., air conditioners, refrigerators, and heaters). Because demand decreases very little in response to higher prices, even small decreases in supply lead to significant price increases. The increase in price can be so large that a single firm that owns several plants can profit from shutting down one plant. Even if its competitors do not cooperate with this strategic behavior, the lost profits resulting from scaling back production in the one plant can be more than offset by the large price increase received for selling power from their other plants (Borenstein, 2001; Joskow, 2001).

These problems caused by insufficient responsiveness to prices were particularly acute in California. Because AB 1890 had frozen retail rates (and AB 265 refroze them for SDG&E customers), consumers were completely shielded from the increases in wholesale prices as the crisis unfolded. They received no signal or incentives to conserve energy and, consequently, power generators were able to bid even higher prices without losing sales.

Finally, because of the importance of real-time reliability, electricity markets are much more sensitive to the market share of large producers. The largest merchant generator in California, AES Corporation, controls less than 10 percent of the total demand on a hot summer day. In most markets, a similarly sized firm has little ability to increase the market price. In the electricity market, though, that 10 percent market share can represent the critical margin of power required to keep the lights on and air conditioners working on days with tight supplies. Consequently, relatively small firms can exercise great influence on prices if their withdrawal from the market will lead to blackouts.

Market power has been a pervasive problem in deregulated electricity markets. Numerous analyses have turned up evidence of the exercise of market power in California as well as in other restructured markets (Wolfram, 1999; Borenstein et al., 2001; Bushnell and Saravia, 2002). In a competitive market, economic theory predicts that firms will sell their goods at a price equal to their short-run marginal costs, the variable cost of producing the last unit produced for the market. Unlike other industries in which firms' costs are not publicly available, these costs are known for electricity generators because of the history of rate regulation and ongoing environmental controls. When comparing actual wholesale market prices to competitive benchmark levels, researchers find observed wholesale prices to be persistently higher. In particular, as demand increases and supplies tighten, generating firms have increasing amounts of market power leading to higher markups over competitive price levels (Borenstein et al., 2001; Bushnell and Saravia, 2002). Although the exercise of market power is common, the California market appears to have been particularly susceptible to manipulation. Firms were able to raise prices above competitive levels even when supplies were comparatively plentiful, and the highest markups occurring during times of tight supply endured for much longer periods.

During the worst months of the crisis—November 2000 through May 2001—there is primae facie evidence supporting market manipulation by generators. To raise prices generators would have had to withhold capacity from the market, and plant unavailability in California was significantly higher when compared to the number of plants off-line before the crisis (see Figure 3.8). In addition, detailed analyses of the costs of generation, taking into account the increased costs of inputs and market conditions, find that about a third of the price increases experienced during the summer of 2000 can he attributed to market power (Joskow, 2001; Joskow and Kahn, 2001a, 2001b).

This evidence, however, is not incontrovertible, and generating firms and other analysts have offered up a number of alternative reasons that prices may have spiked. The market clearing price is determined by a host of specific market conditions, many of which are not publicly available or easily observable (Harvey and Hogan, 2001). Calculations of the competitive market clearing price under competitive conditions are, therefore, crude estimates and cannot by themselves show that market prices were higher than generation costs. High prices may also represent scarcity rents. If the electricity system reaches its capacity with no strategic withholding of supply, supply is essentially fixed. Demanders then will bid up to their maximum willingness to pay to gain access to the fixed supply, potentially driving prices far above production costs.

The evidence concerning plant unavailability is also hotly debated. Generators claim that plants were unavailable because of mechanical failures and not strategic behavior. They claim that their plants were run particularly hard during the 1999–2000 winter, leading to additional failures in the following months. FERC did audit plants to see if they were altering their repair schedules to increases prices (Federal Energy Regulatory Commission, 2001). It found no evidence of such efforts to manipulate plant availability, but the study methodology was later criticized in a review by the General Accounting Office (General Accounting Office, 2001). Also, as Figure 3.8 illustrates, the average amount of capacity off-line has actually increased since the crisis abated in June of 2001.

Alternatively, faulty planning by generators may have led to plant unavailability. Certain plants require long lead times to begin operation, and if generators underestimate the demand for the next day, they may be unable to have the plant up and running in time for the market (Harvey and Hogan, 2001). Finally, certain plants were unavailable because of financial and regulatory chaos. When the utilities lost their creditworthiness in January 2001, they halted payments to QFs that supplied thousands of megawatts. Unable to cover their fuel costs, the QFs were forced to halt production and sued to be freed from their contracts with the utilities. Governor Davis resisted releasing the contracts and allowing the QFs to sell their capacity on the higher-priced spot market because this action would have increased the overall costs of electricity purchased by the Department of Water Resources (DWR). This resistance had the effect of taking 2,000 MW of QF production off-line.

Although there is ample evidence indicating that market manipulation did occur, the legal case that merchant generators unlawfully engaged in anticompetitive behavior remains unresolved. FERC, under the Federal

Power Act, is required to ensure that wholesale generators charge "just and reasonable" prices. During the 1990s, FERC permitted merchant generators to sell their power at competitive rates if they were able to show that the markets in which they operated were reasonably competitive. Although FERC found early in the crisis that wholesale rates were no longer "just and reasonable," it found insufficient evidence to lay the blame on the exercise of market power by generators, and it has yet to decide the extent of refunds, if any, owed California consumers.

The case is complicated by a number of factors. Many of the actions by generators were probably legal, even if they resulted in higher wholesale prices. Even if generator actions were illegal, courts may not intervene because of legal precedents that defer to the rates set by regulatory bodies. Finally, proving market manipulation after the fact places a heavy evidentiary burden on regulatory and enforcement officials. Because of the intricacies of plant operations, it is virtually impossible for an outsider to determine whether a generating plant did not produce electricity for legitimate reasons, such as mechanical breakdowns, or in an effort to increase wholesale prices. Clearly, the California experience has highlighted that identifying and mitigating market power after the fact is politically contentious, administratively burdensome, and legally complex.

Regulatory Missteps

The magnitude of the crisis and the extent of its repercussions were certainly exacerbated by a number of regulatory missteps. The design of California's deregulation has received much criticism. Cambridge Energy Research Associates argues that "partial deregulation" based on a "potpourri of competing stakeholder claims" inevitably led to crisis (Cambridge Energy Research Associates, 2001a, pp. 7–8). The Reason Foundation blames the "chaotic implementation" of deregulation (Kiesling, 2001). FERC points out that California accounted for the majority of problems arising from newly deregulated markets and has deemed California's implementation of AB 1890 as "fatally flawed" (Federal Energy Regulatory Agency, 2000). Hogan characterized California's deregulation as "[a] flawed wholesale market and a caricature of a retail electricity market [that arose] . . . as the product of a volatile combination of bad economic theory and worse political economy practice" (Hogan, 2001b, p. 25). Two problems deserve special attention: the implementation of AB 1890 in which a number of decisions led to excess exposure to spot markets and inaction in the face of the impending crisis.

Excessive Exposure to Spot Markets

Although market conditions and market power help explain the extraordinary run-up in electricity prices, they do not in themselves explain the subsequent financial crisis and collapse of the electricity sector. This aspect of the crisis can be explained only by the utilities' excessive exposure to market risk. The retail rates at which the utilities could sell electricity were frozen, but they were forced to buy that same electricity at fluctuating wholesale rates. In the first years of restructuring, average wholesale rates were well below the level of retail rates. Thus, the rate freeze acted as a rate floor, preventing retail rates from dropping to the low levels of wholesale prices and enabling the utilities to earn additional revenues that they applied toward paying off stranded costs. Despite this early fortunate experience, the utilities were operating with the risk that wholesale rates could increase above these fixed retail levels, leading to losses.

Firms that are exposed to volatile commodity prices typically hedge their risks through forward contracts, long-term contracts, or other financial instruments. For example, it is a common practice for newly deregulated utilities that divest some of their generating capacity to sign long-term contracts in which they buy back all or a portion of the power from the new owner (Borenstein, 2001). California utilities did not sign such contracts and, consequently, controlled less than 70 percent of their total energy sales through owned capacity or long-term contracts. Moreover, at times of peak demand, the utilities owned or had under contract only about 18,000 MW of capacity, only 40 percent of their peak loads of 45,000 MW. Consequently, the utilities were dependent on the spot market, no matter what price was being charged, for the difference between their own capacity and their customer demands.[5]

If California utilities had tied up more supplies through long-term contracts, the volatility would not have had such devastating repercussions. The enormous increases in their energy purchasing costs would have been mitigated, possibly saving them from financial collapse. More important, greater reliance on long-term contracts would have mitigated the exercise of market power, reducing the degree of price volatility. By shrinking the size of the spot market, long-term contracts decrease the benefits of market manipulation because there are fewer megawatts of power that can be sold at high spot market prices.

Neglecting the critical role that long-term contracts play in a well-functioning electricity market was a major failure of the implementation of California's experiment in deregulation. At the time of restructuring, many industry insiders fully understood that commodity markets are inherently volatile and risky. Failure to hedge against these risks was the equivalent of refusing to buy earthquake insurance in California, but the implementation of electricity deregulation, through a series of seemingly unrelated actions, did exactly that.

AB 1890 was mute on the issue of long-term contracting, neither requiring nor forbidding it. Initially, the CPUC implemented restructuring by requiring that all electricity, utility-owned generation as well as nonutility-owned, be bid through the PX spot market. The utilities soon requested that they be given permission to hedge their positions. In a number of decisions dating back to 1999, the CPUC did slowly and at times reluctantly grant the utilities the authority to hedge. In July 1999, the CPUC allowed utilities to buy block forward contracts through the PX for up to a third of their minimum load. Then in March 2000 the CPUC expanded the amount of energy that could be purchased

---

[5]In addition, as is explained below, the CPUC required that all electricity, even electricity generated by the utilities themselves, be sold through the PX. Thus, during the height of the crisis, the utilities were buying electricity from themselves at prevailing spot market prices. To the extent that the utility holding companies retained a portion of the revenues from these high-priced sales of electricity, the losses incurred by their utility subsidiaries increased, exacerbating the financial crisis.

through forward contracts, although it retained the right to disallow contract costs (California State Auditor, 2001a, p. 25). In August of 2000, it again expanded the authority to include bilateral contracts. Despite these moves, the utilities remained inadequately protected from price spikes as the crisis broke.

Part of the explanation for this inaction is that the utilities and regulators were focused on other problems. The early years of deregulation were marked by low prices, and the California Energy Commission, as late as February 2000, was predicting decreasing wholesale prices. Consequently, little attention was being devoted to the risks of price spikes. In the early years of restructuring, regulators were also more concerned about the exercise of market power by the three main utilities rather than the new generators. Because the former vertically integrated utilities controlled the distribution system and vast amounts of generating capacity and had strong customer loyalty, the concern was that they would be able to thwart the development of a fully competitive retail electricity market. The requirement that the utilities divest generating capacity and the restrictions on long-term contracting for power were in large part directed at preventing the utilities from dominating the post-restructuring market.

Efforts to recoup stranded costs also diverted the attention of the utilities and regulators. The utilities wished to recoup these costs as quickly as possible, and regulators wished to end the retail price freeze connected to stranded cost recovery so that consumers could take advantage of what seemed like very low wholesale rates. Requiring that the utilities purchase power through the spot market facilitated this process by simplifying the accounting for these payments (California State Auditor, 2001a; Cambridge Energy Research Associates, 2001a).

A second reason for the lack of action was that neither the major utilities nor the California Public Utilities Commission appear to have grasped how their roles had radically changed in a deregulated market. Deregulation called for customer choice and competition to determine electricity rates. Utilities have claimed, nevertheless, that they fully expected regulators to allow them to recover the full costs of their wholesale power purchases—an expectation more fitting a regulated firm than a competitive one (Edison International Corporation, 2001). At the same time, the California Public Utilities Commission continued to assert the need to review the prudence of the utilities' long-term contracts, ignoring the role that consumer choice and retail competition could play in disciplining bad investments in long-term contracts.

The failure of retail competition to take hold in California further exposed the utilities to unhedged risk. The initial vision of a deregulated market foresaw a market with vibrant competition between numerous retail energy service providers. Within this vision, requiring that the utilities divest all their thermal generating plants and placing restrictions on their long-term contracts was more logical. The new ESPs would need access to wholesale supplies of electricity to serve their customer load, and they were free to use any hedging strategy they found profitable in their wholesale purchases. The utilities, in turn, controlled sufficient generation to serve the customers that remained with them.

Unfortunately, extensive retail competition did not develop. As shown in Figure 3.12, at the height of competitive access in the spring of 2000, about 20 percent of large industrial users, representing over 30

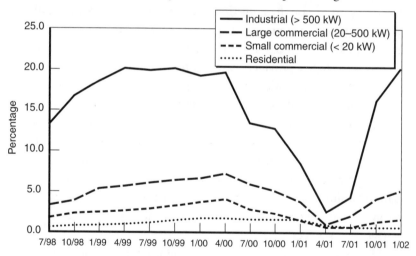

SOURCE: California Public Utilities Commission Direct Access Implementation Activity Reports.

aData for April 2000 were unavailable.

Figure 3.12—Percentage of Direct Access Customers per Fiscal Quarter, 1998–2001

percent of industrial demand, had switched away from utility providers. However, fewer than 2 percent of residential customers and fewer than 5 percent of small commercial customers ever switched. In contrast, in other restructured markets, such as that in the United Kingdom, more consumers exercised choice. Since 1999, 17 percent of residential customers there have opted to buy electricity from a nonutility provider.

This anemic record may be attributed to a number of factors. First, consumers, habituated to receiving steady monopoly electricity, tend to take notice of and comprehend new competitive possibilities slowly. When competitive providers broke into the market for long-distance telephone service, for example, it was ten years before 30 percent of long-distance calls were handled by these new entrants, despite the fact that they offered substantially lower prices. The success of residential choice in the United Kingdom may be attributed to the fact that electricity restructuring and choice for larger industrial firms were in place for many years before they were introduced to residential users.

This customer inertia was reinforced by provisions of AB 1890 that were designed to ease the transition to competition. Residential and small commercial customers were given an automatic 10 percent rate decrease and were protected by the rate freeze. These policies dulled incentives to change providers. New entrants found it difficult to undercut the utilities' prices, and competitors could offer few value-added services to small customers to lure them away from the utilities. The one exception was marketers offering "green power" who attracted environmentally conscious users willing to pay a premium over existing rates. Larger industrial customers, in contrast, benefited more from switching. Because they had not received a 10 percent rate cut from the incumbent utility and because they are heavier users of electricity, new entrants could offer more competitive prices and a host of energy-management services.

The effect of the rate freeze imposed on utilities is evident in the pattern of customer choice. As seen in Figure 3.12, beginning in April 2000 as wholesale prices rose precipitously, customers flocked back to the utilities. Consumers, who had not entered into long-term agreements with new providers to lock in electricity rates, saw their bills soar, and they abandoned competitive ESPs to take advantage of frozen default rates. Pennsylvania experienced a similar collapse of customer choice when wholesale prices rose above the default rate charged by the incumbent utilities. When wholesale rates dropped to pre-crisis levels, larger customers quickly returned to competitive providers although smaller ones did not react as quickly.

Regulators were also ambivalent about retail competition. The CPUC implemented a consumer education program as mandated by the restructuring legislation, but otherwise maintained a hands-off approach (California State Auditor, 2001a). In particular, the CPUC did not mandate programs that promoted the switching of customers to new ESPs. In Pennsylvania, in contrast, where residential choice had been initially more successful, utilities were required to move some of their customers to competitors. Also, the CPUC failed to aggressively pursue a set of interconnection rules for metering and billing that enabled newcomers to enter profitably.

Failure of retail competition was a problem in its own right. Many of the benefits of restructuring, such as cheaper rates, innovative payment options, and energy-management services, were to arise out of the competitive struggle to attract retail customers. But the failure of retail competition had the more immediate effect of increasing utilities' risk exposure. If ESPs had attracted more customers, the utilities would have had to serve less load, decreasing their reliance on the spot market. The CPUC placed the utilities in this bind by working at cross-purposes. It required the utilities to purchase through the spot market in part to promote retail competition, but then it failed to follow through on promoting competition, leading the utilities to serve a larger than expected load through spot market purchases.

Finally, efforts to promote long-term contracting were stymied by mistrust and poor relations between the CPUC and utilities. Although the CPUC did act to permit greater utility use of hedging instruments, it remained suspicious of long-term contracts. It continued to reserve the right to disallow long-term contract costs in the future if spot market prices were below the contract price, and it was slow to review the contracts that were signed. The utilities, for their part, were hesitant to hedge, either because of their mistrust of the CPUC disallowance or because they were overly optimistic concerning future wholesale prices.

During the summer of 2000, for example, the utilities employed little more than half of the forward contracts they were authorized to purchase (California State Auditor, 2001a, p. 26). These decisions turned out to be quite expensive for both the utilities and California in general.

Inaction in the Face of the Impending Crisis

The lack of a rapid, decisive, and coherent response by policymakers also contributed to the devastating consequences of the crisis. The problems with the electricity sector became widely evident in June of 2000 as wholesale prices soared above $100 per MW. The utilities were forced to buy expensive wholesale power and sell it for low retail rates, and their debts mounted rapidly, as much as $50 million per day. The dangers of default by the utilities, widespread chaos, and rolling blackouts loomed large.

During the rest of that year, however, California and federal regulators took only limited actions to address the fundamental problems driving the crisis. In December, FERC issued an order with a set of ultimately ineffective market mitigation measures. In the order, FERC made it clear that it believed that California bore the ultimate responsibility to address the crisis. California took a number of actions. During the summer of 2000, the ISO became concerned about increasingly tight supplies and initiated a program to contract for 3,000 MW of additional peaker plants that could be brought on-line for the summer of 2001 (California State Auditor, 2001b). With AB 265, the legislature reimposed a rate freeze on SDG&E to shield consumers from a run-up in wholesale prices. It also enacted a number of measures to increase supply through the expedited approval of generating capacity and to decrease demand through conservation efforts.

These programs met with moderate success. The ISO was able to contract for 1,324 MW of capacity, although it still required CEC approval for the plants. Expedited review did lead to about 400 MW of peaker capacity being brought on-line during the summer of 2001, and the various conservation programs led to larger-than-expected reductions in demand during 2001. Nevertheless, these efforts failed to avert the financial and institutional cataclysm that hit the California market in January 2001.

More decisive action was gravely complicated by the division of regulatory authority between the state, which regulated retail rates, and FERC, which regulated wholesale rates. Avoiding the imminent financial collapse of the utilities required raising retail prices, placing controls on wholesale prices, or a combination of the two. The political and economic situation at the height of the crisis was complex and full of uncertainty, leading to pitched debate over the correct course of action. There were several competing diagnoses of the fundamental causes of the crisis. These included, among other things, the exercise of market power, a poorly designed deregulatory structure, higher production costs, and scarcity of electricity. Each of these causes led to different policy solutions. To the extent that the crisis reflected higher production costs, and scarcity, price increases were warranted. In contrast, to the degree that high wholesale prices reflected the exercise of market power, price increases would simply ratify the distortions that arose because of that market power, while not solving the underlying problem.[6] Rather, market power called for price caps or other methods of mitigating the exercise of market power.

California politicians and regulators, being closer to California consumers, wished to avoid price increases for their constituents, as evidenced by their support of AB 265. These proclivities were likely reinforced by the extended presidential election in November 2000 as politicians wished to avoid becoming mired in crisis when political advancement in the next administration remained possible. Correctly or incorrectly, they believed that the behavior of the wholesale market was not purely the result of competitive market forces, supporting their decision to maintain retail price controls. They would certainly have reinstituted wholesale price controls if they had had the authority to do so.

FERC, in stark contrast, through its efforts to liberalize electricity markets throughout the United States, was more closely aligned with the power generators with whom it shared pro-restructuring positions. It

---

[6]Increased prices would mitigate the exercise of market power by increasing demand responsiveness, but they would not prevent generators from earning monopoly profits.

advocated strongly that the crisis was not due to the exercise of market power. Interventions in the market would, in its view, risk further damaging the market and impeding the investments in new capacity required to alleviate the crisis in the long run.

Resolving these issues of cause and appropriate policy response would be difficult under any circumstances. Here, the crisis atmosphere, combined with strong ideological differences and mutual mistrust and recrimination between FERC and California, prevented the two from working in concert. Earlier unilateral action by either party—a price increase by California or the imposition of wholesale price caps by FERC—was unlikely because it undermined each party's political and ideological position.

In retrospect, a well-designed combination of increased prices and short-term price controls on wholesale prices implemented in 2000 could have avoided much of the damage of this crisis. They would have protected the utilities from fiscal collapse by simultaneously increasing their revenues and decreasing their costs. Retail price increases would have addressed the fundamental capacity shortage by signaling the need for conservation, and wholesale price caps may have reduced the exercise of market power. Wholesale price controls would have made retail price increases more politically acceptable and retail price increases would have signaled to generators that California wished to maintain a healthy investment environment.

In the end, both California regulators and FERC relented, adopting policies along the lines of this compromise. The CPUC ratified two price increases of historic proportions, one of 10 percent in January and a second larger increase averaging 46 percent in March. At the federal level, the politics surrounding energy regulation changed dramatically during the spring of 2001. President Bush appointed two new members to the Federal Energy Regulatory Commission, and the Democratic Party gained control of the Senate, leading to renewed calls for action. Consequently, FERC switched its antiregulatory stance and imposed effective regional price caps on June 19, 2001. Unfortunately, by the time these policies took effect the damage had been done.

Faulty Market Design

The disastrous performance of the California market has also been linked to the structure of its market institutions. The design of well-functioning electricity markets is intricately complex. Design must incorporate elements of regulation, coordination, and competition. Auctions for power must be designed to provide incentives for least-cost dispatch of power, for the expansion of generation and transmission capacity when needed, and for mitigating the potential abuse of market power. The entire system must be coordinated to control for the externalities of network operations and the need to maintain system balance in real time. The details of market design are critical and getting them wrong can lead to perverse actions by market participants. It is common for deregulated markets to require amendment as design issues arise (Hogan, 2001b).

Several specific deficiencies with the California example have been identified. The overall design was complex, relying much more on market forces than other examples of deregulation. Joskow has characterized it as "the most complicated set of wholesale electricity market institutions ever created on earth and with which there was no real-world experience" (2001, p. 14). During the implementation of AB 1890, energy traders, generators, and other interests bargained over rules paying closer attention to their interests than to efficient and effective market design. In the end, the rules were opaque to all but industry insiders.

An example of this complexity was the separation of the PX, which conducted day-ahead auctions for power, from the ISO that runs the transmission grid and purchases power in real time. This bifurcation of responsibility created incentives to move transactions from the day-ahead market to the real-time market where competition is attenuated because of the exigencies of maintaining system balance (California State Auditor, 2001a; Hogan, 2001a). These bidding strategies complicated the administration of the grid and raised wholesale prices.

Hogan also criticizes the market structure for lacking sufficient pricing zones. Network congestion can cause the market-clearing price

47

to differ from location to location because inexpensive power is unable to flow to areas being served by more expensive generators. California had only a two-zone congestion system that failed to track congestion within each zone. Consequently, price signals for efficient dispatch of generation and for investments in needed additions to transmission capacity were not being properly generated by the market (Hogan, 2001a, p. 27). Revelations of Enron trading strategies have highlighted how faulty market design left the transmission market vulnerable to manipulation. In one strategy, for example, Enron played California's market against regulated transmission in neighboring states. It would claim to ship energy through California counter to the direction of congestion, thereby collecting payments for congestion relief. It would then sell that power back to the original location through regulated transmission in neighboring states. No net energy was moved or congestion relieved, but Enron profited from the spread between California congestion payments and tariffed transmission charges. These problems are particularly important given that transmission bottlenecks have been a major source of concern during the crisis, and additional transmission will be required to improve the operation of the California electricity system.

These market design issues were probably not fundamental factors in the California crisis, certainly not in comparison to the supply shortage and the exposure to the spot market. As Harvey and Hogan admit, "the conditions were so extreme in California that even a good market design may not have survived the summer of 2000 and its aftermath" (Harvey and Hogan, p. 28). Nevertheless, the future of the California experiment in deregulation will depend on getting the details of market design correct.

Conclusion

The tidal wave that struck California's electricity sector from the summer of 2000 through the spring of 2001 was due to a specific combination of factors that befell California. Some common simplifying myths concerning the origins of the crisis do not stand up to scrutiny. It was not due to explosive demand growth in California. Demand was growing more slowly than it had in the 1980s and much more slowly than in neighboring states. Also, it was not caused by rampant "NIMBYism" preventing construction of new generating plants. The regulatory review process slowed plant construction, but new plants were being sited, funded, and built.

Among the remaining factors, no single one fully explains the crisis. The fault cannot be pinned entirely on the shortage in generating capacity. Other states, such as New York, have experienced shortages without catastrophic consequences, and even in California, the worst of the crisis occurred during the winter of 2000–2001, when demand was low. Similarly, market manipulation by generators does not tell the whole story. There is strong evidence of the exercise of market power, but even if wholesale markets had been perfectly competitive, wholesale prices would have increased because of increases in input prices. In addition, blaming market players does not explain why they did not flex their market power to the same degree before May 2000 and after June 2001. The flaws in the restructuring of the electricity sector cannot account for everything. The market, after all, worked reasonably well for the first two years of its operations, and many of the features that have been criticized, such as the retail price freeze, are common to other restructured markets that have performed better.

Although the division of regulatory authority between California and FERC led to catastrophic policy paralysis in response to the crisis, it cannot be blamed for the run-up in wholesale rates that instigated the crisis. Finally, inadequacies in the design of market institutions created greater opportunities for manipulation and impeded coordinated responses to emergency conditions, but such problems were not unique to California. The design of electricity markets is complex, and all efforts at restructuring have encountered unforeseen difficulties, requiring midcourse corrections.

Given the uniqueness of California's experience and the large number of factors at play, it is not possible to fully disentangle the unique contribution of each factor and the interactions between them that led to blackouts, major financial crisis, and the systemic breakdown of market institutions. Some important conclusions can be offered nevertheless.

First, California's electricity sector was rocked by a number of factors unrelated to restructuring: the rise in national natural gas prices, higher costs for pollution permits, and a drought in the Northwest that reduced available imports of electricity. Even if the electricity sector had remained regulated, prices would have increased, and some blackouts would possibly have occurred between May 2000 and June 2001.

Second, market and regulatory conditions aligned, making a particularly ripe environment for the exercise of market power. The shortages in generating capacity played a critical role, increasing the bargaining strength of merchant generators and signaling the enormous profits that could be gained through supply shortages. At the same time, the excessive reliance on the spot market, constraints on transmission capacity, features of the market structure, and the division of regulatory authority all increased the opportunities and incentives for strategic manipulation of the markets.

Third, the exercise of market power fed back to exacerbate underlying problems. It increased the severity of shortages and appeared to interact with the natural gas market, driving up prices to unprecedented levels.

Fourth, increasing wholesale prices combined with the perilous risk exposure of the utilities to create a full-blown financial fiasco.

Finally, the division in regulatory authority and market structure impeded policymakers from developing a rapid, coordinated, and effective response before major damage was inflicted on the electricity sector, the California economy, and all Californians.

An important lesson from the crisis is that electricity systems are complex, interdependent systems. Decisions concerning generation, transmission, distribution, the delivery of energy sources such as gas, and consumption must be coordinated in real time at all times under tight constraints of reliability. The state of competition, market rules, and regulatory oversight interact in multiple and complex ways to coordinate actors, control market power, elicit investments, and send signals to consumers. Changes in certain elements of the system can have profound effects on other elements. Consequently, policymakers must be cautious in dealing with reforms in a piecemeal fashion.

Addendum: The Crisis Fades

Unexpectedly, the summer of 2001 saw the crisis begin to abate. Late in the spring, experts were still predicting ever-higher prices and days of rolling blackouts (Vogel, 2001). Instead, no blackouts occurred and wholesale prices tumbled to their precrisis levels. Although the end of the crisis was unexpected, the underlying reasons for this turn of events are not surprising. Several trends that caused the crisis in the first place were reversed.

The supply shortage abated as movements in both supply and demand brought them more into balance. Available capacity increased for a number of reasons. Over 2,000 MW of additional capacity was brought on-line. Of this capacity, 634 MW had been rushed to market through the emergency fast track regulatory approval process established by Governor Davis, but most of it had been in the pipeline before the beginning of the crisis (California Energy Commission, 2002b). In addition, the amount of unscheduled outages of generating plants plummeted (see Figure 3.8).

These shifts were complemented by lower electricity demand. Even though the summer of 2001 was on average hotter than the summer of 2000, electricity demand decreased noticeably, topping out with an 8.4 percent decrease in June (see Figures 3.13 and 3.14). These conservation efforts were not directly a response to increased retail prices given that they preceded the steep increases that took effect in June. A number of targeted demand-reduction programs, however, offered compensation for reduced usage. For example, the 20/20 program offered consumers a 20 percent rebate on their summer 2001 electricity bill if they reduced their usage by 20 percent. These programs combined with heightened public awareness of the crisis and public appeals for conservation played significant roles in reducing consumption. In addition, price increases for natural gas over the winter, which many consumers confused with increased electricity prices, and a slowing economy appear to have contributed toward curbing demand.

Tight conditions in the markets for inputs for electricity also abated. The price of natural gas tumbled to the single-digit range (see Figure

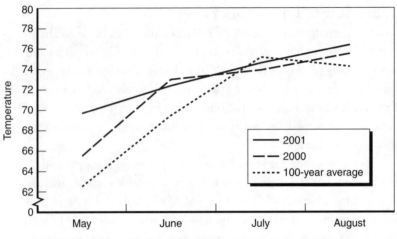

SOURCE: Western Regional Climate Center, http://www.wrcc.dri.edu.

Figure 3.13—California Summer Temperatures, 2000, 2001, and 100-Year Average

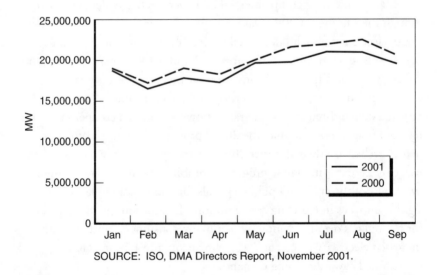

SOURCE: ISO, DMA Directors Report, November 2001.

Figure 3.14—Weather Adjusted Loads, January–September 2000 and 2001

3.9). Electricity generating plants were removed from RECLAIM's NOx trading system and were charged a flat rate for each pound of emissions. These shifts reduced the costs of generation and helped drive wholesale prices down.

Actions taken by policymakers also helped. The price increases implemented by the CPUC not only contributed toward curbing demand, they mitigated the financial turmoil enveloping the electricity sector. It also appears that policy actions that reduced generators' incentives to strategically withhold power decreased the exercise of market power during the summer. First, FERC implemented a strict, regionwide cap on wholesale rates in June, limiting the ability of generators to increase market prices. Second, the Department of Water Resources locked up a large portion of California's energy needs in long-term contracts, thereby reducing the size of the spot market. The decrease in the number of plant outages seems to indicate that these actions were successful.

Since the summer of 2001, the California energy sector has hobbled forward in a muddled, though stable, state. Spot market prices have moderated considerably, hovering near their precrisis levels. Unfortunately, consumers are not benefiting much from these dramatically lower prices, because most power is procured through the long-term contracts signed by the state, leaving only marginal amounts to be bought through the spot market. Hot spells during the summer months of 2002 have shown that California is not yet completely out of the woods. Spot market prices have at times bumped up close to FERC's current $91.87 price cap, and the ISO has declared some emergency notices as reserves have dipped low. The system has held up, nevertheless, and no rolling blackouts have been necessary. The grid has been strengthened by continued additions to generating capacity. Since the beginning of 2001, over 10,000 MW of capacity have been brought on-line or are nearing completion. Nevertheless, the long-run adequacy of California generating plant remains a concern. In reaction to plunging wholesale prices, power producers have either cancelled or delayed almost 5,000 MW of planned construction.

Investigations and court suits concerning allegations of market manipulation have dragged on with no sight of an early conclusion. The most prominent revelation has been the Enron memo detailing the trading schemes it employed to manipulate design flaws in the California market. These schemes focused on ancillary markets that are much smaller than the main market for wholesale power, and they were not responsible for the overwhelming problems faced by California. Despite providing no smoking gun for the more fundamental allegations of market manipulation, the memo has given credibility to the charges and has prompted much closer scrutiny to the practices of energy traders.

California, however, cannot allow its electricity management to continue to drift, as significant challenges lie ahead. FERC's market mitigation measures expired at the end of September 2002. It declared its intention to replace the current price cap formula with a straight $250 per MW cap, and California regulators fear that this increased level may reopen the market to manipulation. Also, the state's authorization to purchase power expires at the end of the year, requiring that the three main utilities resume their traditional role of purchasing power for their customers. Most important, California continues to lack a clear direction out from the crisis. The next chapter reviews options for moving forward, and Chapter 5 examines three specific recommendations for improving the performance of the electricity sector.

4. Rebuilding the California Electricity Sector: Institutional Choices

To this point in the California electricity crisis, policy action has focused on the financial dimensions of the crisis: paying for very high-cost wholesale power, rescuing the utilities from insolvency, seeking refunds for unjust and unreasonable wholesale rates, and financing the state's $42 billion in long-term contracts. These problems remain unresolved and must be addressed before progress on longer-term issues can be made.[1] As these financial problems are addressed, long-run, institutional issues will become prominent.

The electricity crisis shattered the main structures of California's restructured market. The PX went out of business, competitive electricity retailers disappeared, the state replaced the utilities as the main purchaser of wholesale power, and the independent, stakeholder board of the ISO was replaced by a board appointed by Governor Davis. A host of new state and federal regulations were enacted in a frenzied response

---

[1] The Department of Water Resources authority to purchase electricity expires at the end of 2002. At that time, the utilities must be financially able to resume their role of purchasing power for their customers. SCE and SDG&E are well along toward recovery. The fate of PG&E, in contrast, remains less certain, hinging on the decisions of the bankruptcy court. PG&E and the CPUC have offered competing plans for resolving creditors' claims, and decisions could lead to protracted legal wrangling. The DWR's purchasing authority may need to be extended for a short period, but the additional purchases necessary would be small because the state has already locked up most power needs with its long-run contracts.

Settling the payments for the expensive long-run contracts has been more contentious. It has already led to a prolonged dispute between the CPUC and the DWR over ratemaking authority and to the curtailment of retail choice. A bond issue to repay the general fund for power purchases was repeatedly delayed because of these disputes. Continued battles attempting to shift these costs to different groups of ratepayers and taxpayers can be expected.

to the crisis, and confidence in electricity markets was battered by the crisis and subsequent revelations of market gaming by energy traders. This list could be extended.

Policymakers face numerous fundamental questions as to how to repair and replace these fractured structures. How will the electricity sector be organized? What segments of the industry, if any, should be open to competition? What rules will dictate the competitive process? What role should the state take on as regulator, planner, and direct participant in the electricity sector? What role should the various state agencies play? Although significantly less attention has been given to these concerns, how they are resolved will determine whether the crisis becomes a long-term drag on California's economy or whether California can put the crisis behind it and build a healthy electricity sector that can fuel future economic growth.

Decisions over the long-run institutional structure of California's electricity sector are complicated by the complexity of the issues that the crisis unearthed and the wide range of options being debated. Serious proposals representing almost the entire spectrum of economic philosophies are receiving significant attention. These include calls for increased public ownership of the electricity sector, a return to the system of cost-of-service regulation that preceded California's experiment with markets, continuing with market reforms by repairing market mechanisms and reinstituting customer choice, and hybrids that combine elements of two or more of these options. These alternatives propose starkly different approaches to the interrelated problems identified above—shortages in generating capacity, the exercise of market power, and weaknesses in regulatory and market institutions.

Assessing the relative performance of these disparate alternatives poses significant analytical challenges. The policy sciences have a number of theories that detail potential government failures in this area. Government-owned and government-regulated industries tend to face dull and at times perverse incentives that limit their ability to use resources efficiently. For example, government bureaus are often internally motivated to increase their size and scope in an effort to increase their organizational prestige and influence, but these internal incentives unchecked by competition may run counter to the interests of the customers they serve. A large literature on the regulation of industry has also shown that regulatory agencies tend to be captured by the industries they are intended to regulate, serving their interests instead of the interests of consumers. In addition, regulatory powers are often used to advance narrow political goals by rewarding certain consumer groups at the expense of others.

Similarly, the policy sciences have developed a refined understanding of market failures. These include the dangers of market power when there is insufficient competition between suppliers; the costs of externalities, such as environmental damage, that are not accounted for in market transactions; and the difficulties of coordinating complex activities in real time through market-like processes.

Despite these important insights, the policy sciences have made much less headway in their ability to assess the *relative* costs of government failures versus market failures in complex situations. In the electricity sector, for example, experience has shown that both government failures and market failures are manifest. Consequently, measuring the performance differences among imperfect government ownership, imperfect regulation of private monopolies, and imperfect market competition is difficult, tradeoffs are harder to assess, and conclusions are ambiguous. As a result, the debate over institutional alternatives is often driven more by ideological predispositions than by hard evidence. Proposals often focus on isolated facets of the electricity sector and leave their normative motivations implicit, muddying the differences between proposals.

To focus and clarify our analysis, we begin by enumerating six basic goals for a well-designed electricity system. A clear set of goals provides a more consistent basis for the comparison of alternatives, helps distinguish between what is known and what remains ambiguous, and aids in identifying the tradeoffs posed by these alternatives. This chapter then applies these goals to understand the broad institutional choices facing policymakers.

Goals for the Electricity Sector

A well-designed electricity sector, whether competitive, regulated, or some form of hybrid, should seek to achieve six goals:

- **Low prices.** The quest to lower electricity rates had been a significant, if not the primary, motivation behind restructuring. Because electricity is a necessary input to almost all aspects of a modern economy, lower prices improve the economic competitiveness of California businesses and benefit California consumers.
- **Bill stability.** Because most consumers cannot easily change their consumption habits in the short run, they seek to avoid spikes in prices. Firms, for example, typically enter into long-term contracts for commodities that provided relatively predictable charges over time. Bill stability is not the same goal as price stability. Prices may be allowed to fluctuate in accordance with short-run market conditions while average monthly bills remain relatively constant (Friedman and Weare, 1993; Borenstein, 2001).
- **Efficient resource use.** Production efficiency requires that all electricity plants be run optimally, that lower-cost sources of electricity be employed before higher-cost sources, and that over time investments in capacity track trends in demand. Consumption efficiency requires that electricity be employed in its most valued uses and that it be conserved whenever the cost of electricity exceeds either the benefits derived from electricity or the costs of efficiency-enhancing investments.
- **System reliability.** Given the high costs of blackouts, at least for the majority of consumers, the reliability of the electricity system has been a major concern. One-hundred percent reliability for all customers at all times, however, is not the ultimate goal. Certain customers place a lower value on an uninterrupted supply, and they can be provided with interruptible contracts while preserving a reliable supply for the majority of users.
- **Administrative feasibility.** Regulatory and market institutions must provide producers and consumers with clear and stable rules that establish incentives and opportunities for economically rational actions. These institutions must possess the authority and ability to perform assigned tasks. At the same time, these agencies must be sufficiently flexible to react to changing

circumstances as the electricity sector continues to evolve
rapidly.
- **Environmental protections.** All these goals also need to be
achieved within the constraints of preserving clean air and clean
water and protecting key environmental resources.

These goals, of course, involve tradeoffs. Maintaining low prices may
harm system reliability by decreasing incentives for investments in
generating capacity and may impede conservation efforts by reducing the
incentives to manage consumption. Some policies aimed at enhancing
the efficient use of resources can be complex, straining administrative
feasibility. Maintaining system reliability may require additional
generation plants that cause environmental damage. Providing
consumers with stable bills may impede the efficient use of resources by
dampening the incentives faced by producers and consumers.
Alternatives differ not only in how well they achieve these six goals but
also in the tradeoffs between goals they entail.

Public Power

Advocates of the public ownership of the electricity system received a
boost from the California electricity crisis. Municipal utilities, especially
those that owned generating capacity such as the Los Angeles
Department of Water and Power and the Sacramento Municipal Utility
District, weathered the crisis in excellent shape. Since the beginning of
the crisis, a variety of proposals to increase public control have been
floated. The creation of the California Consumer Power and
Conservation Financing Authority (CPA) is the most prominent effort,
but it is only one of many. In early efforts to rescue the utilities from
insolvency, the state considered buying their transmission assets. More
recently, Assemblyman Keeley has floated the idea of buying out PG&E
and running it as a public service corporation. Municipalities have also
shown great interest in assuming a role in electricity provision. This
trend is highlighted by the recent vote in San Francisco that narrowly
failed to create a new municipal power authority.

Public power already plays a substantial role in the electricity sector.
More than a fifth of all power in the United States is provided by public

entities. The Energy Information Administration reports that 8.2 percent is produced by the federal government and 14.7 percent is produced by publicly owned utilities and cooperatives. In California, the proportion is even higher, with about 25 percent of power coming from municipal utilities and cooperatives.

Because public utilities have operated alongside regulated, private utilities for many years, there has been extensive research comparing their performance. The data comparing both public and private, regulated utilities with competitive utilities are sparser because restructured electricity markets are relatively new. There are reasons to expect that ownership would lead to different outcomes because public versus private utilities face divergent incentives, opportunities, and constraints. Private utilities face stronger incentives to economize because private owners capture the benefits of costs savings and innovation. These incentives, however, are dulled by regulation, and the benefits of economizing can be dissipated if regulators are captured by the utilities they are intended to regulate. Public ownership, in contrast, avoids the informational costs associated with regulation, but at the same time introduces potential inefficiencies because of the constraints of managing within a bureaucracy and the lack of strong incentives to economize.

Despite these differences, comparisons between the costs of public and private, regulated utilities do not establish an overwhelming advantage for either form of ownership. As shown in Table 4.1, during the 1990s in California, the average cost of municipal power was about 13 percent lower than the average of the three utilities, 8.29 compared to 9.53 cents per kWh for the three major utilities. Much of this difference, however, is due to specific advantages provided to municipal utilities that are not available to investor-owned utilities. The municipal utilities are granted access to low-cost hydro generation from federal projects, are exempt from state and federal taxes, and are 100 percent debt financed, which is less expensive than financing with a mix of debt and equities. In addition, the cost performance of municipal utilities in California is highly variable, indicating that government ownership by no means ensures lower overall costs.

More comprehensive analyses of the relative performance of public and private utilities also arrive at mixed results, although they provide

Table 4.1

System Average Rates (cents/kWh)

	Average, 1990–1999	Standard Deviation, 1990–1999	Standard Deviation Rank
Merced Irrigation District	1.41	0.433	5
City of Vernon	5.68	0.885	18
Santa Clara Municipal Electric Department	5.83	2.382	24
City of Palo Alto	6.13	0.932	20
Modesto Irrigation District	6.31	0.565	9
City of Redding	6.35	0.657	12
Turlock Irrigation District	7.08	0.609	10
Sacramento Municipal Utility District	7.51	0.248	2
Imperial Irrigation District	7.71	0.823	17
Roseville Electric Department	7.97	0.242	1
City of Azusa	8.06	1.697	23
City of Anaheim	8.53	0.735	13
Los Angeles Department of Water & Power	8.69	0.421	4
City of Glendale	8.76	0.781	16
Colton Electric Utility Department	8.88	1.461	22
City of Pasadena	8.93	0.774	15
San Diego Gas & Electric	9.04	0.453	6
Burbank Public Service Department	9.19	0.609	11
City of Lodi	9.39	0.895	19
City of Alameda	9.60	0.773	14
Pacific Gas & Electric	9.75	0.517	8
Southern California Edison	9.79	0.407	3
City of Riverside	9.86	0.501	7
Lassen Municipal Utility District	10.23	1.150	21
Average of three private utilities	9.53		
Average of municipal utilities	8.29		

SOURCE: Bay Area Economic Forum (2001c).

some support for a small cost advantage for public utilities. A recent survey found six studies that concluded that public electricity providers had lower costs than private, regulated utilities; five studies that found no difference; and two studies that found that private, regulated provision led to lower costs (Kumbhakar and Hjalmarsson, 1998). Another recent study echoed these conflicting findings, concluding that public utilities have lower overall costs even accounting for public subsidies but that private utilities are more efficient at generation (Kwoka, 1996). The same study also found that public utilities tended to have slightly lower rates for residential customers balanced by slightly higher rates for commercial and industrial customers, indicating that public managers do exercise discretion in rate decisions.

Public utilities in California, however, have not provided their customers with particularly stable rates. The standard deviation of rates—a measure of variability over time—for the private utilities are among the lowest, and public utilities, such as Burbank, Lodi, and Alameda, that have average rates similar to private utilities have much more variable rates. Most likely, public utilities are more vulnerable to external shocks such as changes in fuel prices or interest rates because they are smaller than the private ones.

Concerning reliability and environmental effects, public ownership can have positive attributes. Public providers may be better prepared to invest in expanded capacity as demand increases because they are insulated from the vicissitudes of financial markets and strict profit constraints. Freedom from the need to maximize profits also enables public utilities to pursue environmental goals. For example, the California Power Authority has decided to focus its investment portfolio on green power sources. There is no strong evidence, however, indicating that public providers are able to deliver these benefits in practice.

In a transition to increasing the role of public power in California, policymakers would have to navigate several hurdles. A number of changes in the power industry make it increasingly difficult for public utilities to provide power on more favorable terms than private utilities, especially if they must shoulder a portion of the added costs of the state's long-term power contracts. New public providers are unlikely to gain access to cheap federal power, forcing them to either construct new generation plants or buy electricity on the open market. Regulatory rules governing access to tax-exempt bonds are in flux and could increase the financing costs faced by new public utilities.

In areas that are currently being served by private utilities, municipalization efforts face stiff resistance from the incumbents who wish to maintain their customer base. Acquisition of utility assets typically entails a long political and legal battle, demanding a high degree of perseverance on the part of policymakers. Most important, the value

of the assets must be negotiated. Recently, electricity system assets have
been trading at above book value (e.g., original cost minus accumulated
depreciation), and as this premium above book increases it becomes
increasingly difficult for a new public power provider to offer attractively
priced electricity.

Return to Regulation

A return to a fully regulated industry remains a possibility and has a
number of influential advocates within regulatory agencies and the
legislature. In April 2002, the Public Utilities Commission reimposed
cost-of-service regulation on the three main utilities, although this ruling
applies only to the assets that they continue to control. This action
could be a temporary stopgap until other policies are formulated or it
could be the first step in reconstituting the old regulatory regime.
Beyond California, the situation is in flux. FERC, some states, and some
foreign countries continue to pursue electricity sector restructuring
aggressively. Nevertheless, the majority of states still have not
restructured, and a number have backed off or slowed down their reform
efforts since the California crisis.

The main benefits of the regulated regime were that it provided a
high degree of system reliability and bill stability. Regulated utilities had
the obligation to plan and construct capacity to service all ratepayers in
their regions. The CPUC in turn set rates that allowed them to earn a
fair return on all capital investments that were used and useful. This
regulatory compact had a long and successful record in stimulating
investments to meet growing electricity demand, and with cost-based
rates it protected consumers from dramatic changes in their bills. The
California electricity shortage of 1948, however, is an important
reminder that regulated systems are not immune from short-term energy
crises. The sets of rules and procedures that had been developed over
time provided a transparent, well-understood, and administratively
manageable process for regulatory decisionmaking. More recently, a
number of successful conservation programs were implemented that
encouraged regulated utilities to search out and invest in conservation.

The regulated regime, nevertheless, suffered from distinct and well-
known disadvantages. Although regulated utilities had historically

provided fairly low-cost electricity, the high rates paid by California consumers before restructuring clearly demonstrate that regulation does not ensure low rates. Utilities operating under cost-of-service regulation are guaranteed to recoup their reasonable expenses plus a reasonable profit on prudent investments. In practice, standards of reasonableness and prudence leave great discretion with the utilities. Consequently, they are at best weakly constrained to operate efficiently, innovate, and develop the lowest-cost portfolio of generation and other assets. The obligation to serve, in addition, creates a bias toward excess capacity, which was one of the root causes of the high rates experienced in California in the early 1990s.

Historically, regulators have failed to provide consumers with price signals that reflect the underlying costs of the electricity they consume. They tended to charge consumers the average cost of all electricity produced. Thus, prices were high when a system had an overabundance of capacity driving up average costs, and prices were low when there was too little capacity and overall costs were lower. As a result, consumers received perverse incentives. When supplies were tight making conservation necessary, prices were low, encouraging greater use. When supplies were abundant the exact opposite occurred. Instead of setting low prices to encourage consumers to take advantage of the available power, regulators set prices high. California's deregulatory experience repeated this trend. Price freezes prevented retail rates from jumping as shortages led to wholesale price increases in 2000 and early 2001. Now, as the shortage has waned, helping wholesale rates decrease substantially, consumers are paying about 50 percent higher rates since the two rate increases implemented in 2001.

Regulators can mitigate these regulatory constraints. Strict cost-of-service regulation can be replaced by performance-based regulatory regimes. These regimes provide utilities greater incentives to operate efficiently by allowing them to retain a portion of the costs savings from their actions. Ratemaking can also be reformed to provide consumers better incentives by aligning rates more closely with the real costs of producing electricity (e.g., the short-run marginal cost) or by real-time pricing (RTP), which is discussed in greater detail below. Although there has been some experimentation with these reforms, regulators have been slow to adopt them, largely because they impair the transparency of traditional cost-of-service regulation and because they fear that certain customer groups may be harmed by changes in ratemaking procedures.

It may be impossible, or at a minimum politically difficult, to put the genie back into the bottle and return completely to the prerestructuring environment dominated by three vertically integrated utilities. A move to return the generation capacity sold off to merchant generators during restructuring to the utilities would be resisted by both the merchant generators and the utilities that are working to move away from their regulated businesses.

A return to a more regulated regime, however, could be achieved without disturbing the current ownership of generation capacity. Retail regulation could be retained while mandating that the utilities procure additional energy through long-term contracts. Regulation of energy procured through long-term contracts is not new. Even in 1995, California utilities generated only about 58 percent of the power while they served three-quarters of the load, indicating that much of their power was purchased from third parties, mostly QFs or imports. Regulators can simply build on this model.

Determining the reasonable and prudent costs of a portfolio of long-term contracts does pose difficulties. If utilities are not exposed to retail competition, regulators cannot eliminate all prudence reviews. In that case, utilities retain market power over their captured customers and have obvious opportunities for self-dealing or signing sweetheart contracts at above-market rates.

At the same time, strict prudence guidelines that view with suspicion all long-term contracts struck at prices above realized spot market prices place an unrealistic burden on utilities. Utilities develop a portfolio of forward and spot market power in an environment of great uncertainty over future market trends. They base their decisions on their expectations of future prices and their need to hedge risks, but as spot and forward prices for electricity vary over time, decisions that appeared reasonable ex ante may appear less so after the fact. Although forward contract prices should roughly track spot market prices *on average*, one must expect that forward contracts, entered into reasonably and prudently, will at times be priced below the spot market price and at other times be priced above. The mere fact that forward contracts are above spot market prices at a particular time, therefore, does not indicate that they were imprudent.

The degree of prudence review to apply to such a portfolio was a major bone of contention between the utilities and regulators leading up to the crisis. The fact that they were unable to resolve these disagreements was a major contributor to the crisis and cautions about the administrative feasibility of regulating such contracts. A return to a more regulated regime, consequently, would require that regulators and utilities strike a compromise. Regulators must maintain a level of review that guards against opportunistic behavior, but such prudence review must be sufficiently limited in scope as to provide utilities with sufficient predictability that they will enter into long-term contracts.

Fix Markets and Proceed with Restructuring

The third institutional option open to decisionmakers is to continue with the restructured electricity market while correcting those components of its design and implementation that led to the crisis. This option receives strongest support from economists and business interests. It also best accords with the policy direction of the Federal Energy Regulatory Commission as it works to establish regional electricity markets in the United States. In the long run, this path presents the largest potential benefits in terms of greater efficiencies and lower prices for consumers. At this time, when California consumers and businesses face extraordinarily high electricity rates for many years to come, the benefits from wringing additional efficiencies are more important than ever. This avenue, however, also poses the most unknowns and the greatest risks. It poses significant administrative difficulties for both policymakers who are faced with a complex set of policy decisions and for consumers who are required to become much more active in their usage decisions.

In other industries such as airlines, trucking, and railroads, reforms to rely on competition instead of economic regulation to manage industry performance have been important policy successes. Deregulation has led to lower prices, more efficient operations, and expanded consumer choice.[2] These success stories have been a main factor motivating experiments with competition in the electricity sector, and there are strong reasons to believe that these benefits can be extended to the electricity sector. Competition can force firms to make wiser investment decisions and operate their plants more efficiently. Some studies of the early results of competition show that it can lower prices and costs (Kwoka, 1996; Bay Area Economic Forum, 2001b). On the other hand, the large body of evidence indicating that privately owned generation has not consistently operated more cost-effectively than publicly owned generation suggests that the magnitude of the short-run benefits from improved plant operations is not overwhelming. Benefits from competitive generation are more likely to accrue over the long run as competitive firms develop lower-cost portfolios of base load and peaker plants and retire older less-efficient plants. As long as the industry is sufficiently competitive, these lower costs will result in lower prices benefiting consumers.

Competition can also spur electricity service providers to develop innovative service packages to benefit consumers. Under the regulated regime, utilities offered a limited range of price schemes and service options, but there is little reason to believe that this one-size-fits-all policy is best for all consumers. The cellular telephone industry, for example, offers a whole range of price and service options that enable consumers to pick a package that best fits their usage patterns. Because regulated prices do not fluctuate with underlying costs, they provide consumers with predictable and stable bills, but they also fail to reward consumers for conserving during times of high electricity prices. Some customers would prefer a service package that offered less stability but helped them to reduce their average bills and manage their consumption

---

[2]Of course, there are critics of deregulation who claim that it has not always lowered prices and that lower prices have come at the expense of poorer service quality. Nevertheless, if consumers receive similar price and quality options under competitive and regulatory regimes, there still are reasons to favor deregulation. It reduces the need for government bureaucracy and it empowers consumer choice in that they can discipline poor performance by taking their business elsewhere, which is easier than complaining to a regulator.

more intelligently. Others who preferred stability could retain service options that more closely matched the regulated rates.

Beyond price options, competitive providers could provide new energy services. For larger users, they could market energy management services, and for residential users, they could offer options such as green power—electricity generated by environmentally friendly sources and a popular offering before the crisis. As such services were deployed, they would also have the beneficial effect of spurring demand for and innovation of new energy-management technologies and energy-efficient appliances.

Securing these potential benefits from competitive electricity markets, however, is not inevitable. It requires that both regulators and consumers successfully address the complexities and challenges posed by competitive electricity markets. Other industries that were successfully deregulated, such as airlines and trucking, were structurally competitive. To move toward competition, regulators had to do little more than abandon entry and price controls.

Electricity poses more entrenched issues. Transmission and distribution, two critical components of the electrical system, remain monopolized, requiring regulatory action to ensure access for competitive generators and electricity service providers. Reliable operation of the grid entails balancing the input, output, and flow of electricity at all times, requiring close coordination between all actors. Consequently, the development of a competitive electricity market involves significantly more complex market design problems than previous deregulatory efforts. California has already learned the hard lesson that improper design and implementation can be disastrous. The California debacle, however, remains unique among restructured electricity markets, and there exist numerous more successful reform efforts on which California can and should model its markets.

Successful competition also requires that consumers of electricity become more aware and active. To this point, they have little such experience because under the regulated regime they enjoyed stable and simple bills. To take advantage of the opportunities provided by deregulation, they would have to understand their consumption habits better and be able to assess the implications of differing service options.

One should not underestimate consumers' ability to make such decisions. After all, they make similarly demanding decisions concerning airline tickets, mortgages, and retirement accounts. At the same time, one should not overestimate consumers either. Given that educated and informed policymakers badly misjudged California electricity markets, it can be expected that numerous consumers would similarly make poor choices if they were confronted with these new consumption decisions.

To move forward with restructuring, California must, at a minimum, redesign its market in a manner that avoids the worst errors that led to the crisis: allowing reserve margins to fall dangerously low, creating an environment conducive to the exercise of market power, market rules that were easily manipulated, a regulatory structure that impeded decisive policy action, and an overreliance on the spot market. Until credible policies are put into place that effectively address these problems, state politicians and voters are unlikely to support a rapid return to competitive electricity markets. In a PPIC Statewide Survey conducted in February 2002, well after the peak of the crisis had passed, only 23 percent of voters supported further deregulation. Thus, if California is to continue down the path of restructuring, it must implement major reforms to ensure that its markets are competitive and will benefit consumers. These include the development of an infrastructure policy to ensure adequate supply, new policies to control market power, a redesign of the market, and the reestablishment of retail competition.

Infrastructure Policy

California must develop a healthy investment environment in which private investors build sufficient new generating capacity to meet the state's growing demand for power. Adequate supplies are essential for maintaining system reliability and avoiding the huge price spikes that hobbled California. To accomplish this goal, policymakers must design a market that provides firms clear signals when additional supplies are required and enables them to bring new supplies onto the market in a timely manner.

In the implementation of California's electricity deregulation, increases in spot market prices were the only signal provided to generators that new investments were required. In this way, California relied on the most highly decentralized, market-oriented system possible to manage long-run electricity supply. In contrast, market designs implemented in other areas incorporate differing levels of regulatory oversight. Four models can be identified. Ranging from the most market oriented to the most centrally planned, they include: (1) reliance on spot markets as in California, (2) markets for capacity, (3) direct payments for capacity, and (4) state planning for capacity. The main tradeoff presented by these options is between the greater efficiency of the more market-oriented proposals and the more stable prices offered by the options that include a great degree of centralized planning.

Although the current system of relying on spot market prices to induce investment failed in California, it has performed well in other deregulated markets. In Chile and Victoria, Canada, for example, reductions in reserve margins did lead to increases in wholesale prices, although not the sudden and sharp price spikes experienced in California. Higher prices induced timely expansions in capacity, which then led wholesale rates to decrease smoothly (Bay Area Economic Forum, 2001b). Relying fully on spot markets offers the prospect of greater long-run efficiency because over time market participants could learn what is the best level of reserve capacity and adjust their actions accordingly. Spot markets, however, are volatile. In California, spot market prices remained low as reserves tightened and then spiked suddenly when system capacity was being strained. Such volatility complicates long-run planning for investors. More important, for such a market to perform successfully, consumers will need to be willing to curtail energy use in response to tightening market conditions.

A second model, implemented in New York, New England, and the PJM Interconnect operating in Pennsylvania, New Jersey, Delaware, and Maryland, develops a market for electricity capacity. A regulator or other market-coordinating body mandates that all electricity service providers maintain control over sufficient capacity to serve their client base and provide a preset level of reserves. This capacity can be provided by owning physical generation plant, contracting with other generators, or purchasing capacity rights on a spot market. In this capacity market, when the price of purchasing capacity rights exceeds the cost of constructing additional capacity, generators have incentives to build their own capacity or contract for new capacity. Some experts argue that a capacity market provides investors better price signals that will increase more gradually as reserves tighten and that are easier to understand (Bay Area Economic Forum, 2001b; Cambridge Energy Research Associates, 2001a). The advantage of enabling the regulator to set reserve levels explicitly is that it can promote price stability by keeping margins sufficiently high. The disadvantage is that the regulator may choose an inefficiently high level of reserves, forcing firms to have a higher cost structure than necessary.

A third model calls for the regulator to provide incentive payments to generators when they make capacity available. In Argentina, for example, all generators that sell power during periods of peak demand receive a fixed payment for the capacity they made available, and generators that offered capacity that is not used receive a variable payment. As with the capacity market model, this system provides direct and clear incentives for providing capacity to the market. Its main disadvantage is that the regulator must set the payment level, a task that is more difficult than establishing a capacity reserve requirement. Payments that are too high allow generators to earn above-market returns, and payments that are too low may fail to provide sufficient incentives to expand capacity when needed. This system has functioned well in Argentina; the United Kingdom recently abandoned it in favor of a system similar to California's spot market approach.

In the fourth model, California can manage infrastructure planning and investments more directly. Before restructuring, California practiced integrated resource planning, an effort to combine the forecasting of energy demand and capacity needs with the development of plans to meet energy needs in the most cost-effective manner possible. Under restructuring, the role of integrated resource planning diminished, but the state could once again undertake a more central role in long-run planning that either augments or works in conjunction with private sector investment decisions. Active state-level planning can contribute to the stability of the system by alerting policymakers to impending shortages, but such planning is no panacea. State forecasters, after all, failed to foresee the shortage that hit the state in the summer of 2000.

The newly created California Power Authority, in consultation with the CEC, recently developed such a broad investment plan, although there is no requirement to update the plan in future years. The California Power Authority has also been granted broad authority to expand on these planning duties by directly owning or investing in power plants. A public role in investment planning could counter the biases of market-driven investment. Incumbent generators enjoy higher prices and profits when supplies are tight, potentially creating a bias against investment. In addition, private investors are beholden to the unpredictability of capital markets and, since the Enron bankruptcy, investment funds have dried up, forcing the cancellation or delay of several projects. The main danger of public investment is the potential for crowding out. If the CPA consistently builds as reserves drop, wholesale prices may remain depressed, deterring private power producers from investing in California. A preferable alternative would be for the CPA to assist the private sector with project funding, thereby avoiding the dangers of crowding out while helping ensure the construction of necessary capacity expansion.

Under all these models for signaling the need for capacity expansion, policymakers must also ensure that power projects can move smoothly and with a minimum of delay from identifying needs for additional capacity, to procuring regulatory approvals and financing, to construction and startup. The California crisis clearly showed the consequential effect that short-run shortfalls in generating capacity can cause. Policies that facilitate the development of new supplies in a timely manner, therefore, are needed. Policymakers have several options that may ease the financing, regulatory review, and construction processes.

Expanding the role of long-term contracts for wholesale power is an important first step. Permitting investors to sell their power in forward markets or through bilateral contracts will enable them to get firm price commitments before plant construction, thereby easing the financing of capacity additions. The permitting process is in need of streamlining. Since 1998, there has been a strong supply of applications for new plant construction in California, indicating that the regulatory approval process has not deterred investment. Nevertheless, the length of the process remains a concern. Fewer delays would shorten generators' planning horizon, making it less likely that unexpected spurts in demand would outstrip supply while a facility is under construction.

The state and in particular the California Energy Commission undertook several actions during the crisis to expedite the review of new proposals. Many peaker plants were constructed under these guidelines and expedited review options remain in force. Given that California's electricity demand will continue to grow, requiring additional plants, these improved procedures must be maintained and strengthened. If expedited review diminishes environmental checks, there is a risk that environmental goals may be compromised. Further study on how these actions affect the tradeoff between expanding capacity versus preserving environmental and community quality will be needed.

Controlling Market Power

Vigorous competition is needed if consumers are to benefit from restructured electricity markets. Ensuring that electricity markets are workably competitive requires a multipronged strategy including measures on how the market is regulated and how it operates. On the regulatory side, California needs to reassess its relationship with FERC, establish price caps, develop policies that prevent the strategic withholding of generating capacity from the market, and reconsider the structure of the electricity sector. On the market side, it needs to ensure that there are adequate reserves and improve the demand-side responsiveness in the market.

One consequence of AB 1890 is that it split regulatory authority between FERC, which now regulates wholesale markets, and the CPUC, which retains control over retail markets. During the height of the crisis, this divided authority fomented strident differences between state and federal regulators, impeding policy action. To avoid repeating that policy failure, California policymakers need to understand FERC intentions and cope with the interdependencies between state and federal decisions.

FERC is statutorily mandated to ensure "just and reasonable" wholesale rates. Early in the crisis it determined that wholesale rates were not "just and reasonable," but it declined to intervene aggressively, leaving California to address the crisis alone. FERC did change directions later, imposing binding, regional price caps in June 2001.

That action paired with its more recent response to revelations of market gaming by Enron appear to signal a renewed seriousness concerning market policing, but other actions have sent mixed signals. FERC replaced its existing price cap, formerly set at $91.87, with a much higher $250 cap, and FERC has been slow to act on the multiple complaints of market mitigation that California has brought.

If California can have confidence that FERC will be a more vigilant market watchdog in the future, the state has greater flexibility in the way it transitions back to competitive markets. For example, it can pursue policies that promote goals such as consumer protection with the confidence that even if these policies impinge on market competitiveness, FERC can act as a backup in case the forces of competition slacken excessively. If, on the other hand, FERC retains its laissez-faire ways, California faces more constrained policy choices. It must develop a regulatory and market framework that ensures that market forces operate strongly at all times, even at the cost of neglecting other goals such as bill stability, administrative feasibility, and environmental concerns. Such a competitive market would involve, among other features, fully exposing consumers to the volatility of wholesale market prices and ensuring that new capacity can be brought onto the market quickly. To do otherwise risks repeating the unchecked exercise of market power. (See Wolak, 2001, for an excellent discussion of market design in the absence of FERC control of market power.)

FERC also continues its attempts to expand its authority over electricity markets by striving to organize a small number of regional energy markets with common market rules within the United States. This effort is highly controversial and is being contested in the courts and in the U.S. Congress. If FERC prevails, however, California may have no choice but to work toward reestablishing competitive wholesale markets and remold its regulatory goals and methods accordingly.

Although no panacea, price caps are an important tool for controlling market power. The debate over price caps was one of the most highly politicized dimensions of the California crisis. Advocates, including many California officials, argued that price caps were necessary to protect ratepayers from gouging by generation firms. They saw caps as the most direct and powerful policy tool for accomplishing this goal.

Free marketers, including key FERC commissioners, retorted that price caps do nothing to ameliorate the fundamental market imbalances causing high prices and lead to inevitable market distortions worse than the original problem.

The main criticism against price caps leveled by economists is that they lead to market shortages by discouraging investment in additional capacity. Almost everyone who has taken a course in introductory economics has been exposed to the argument that rent control (i.e., price caps on rental housing prices) encourages landlords to remove units from the market, leading to a shortage. A critical point largely missing from the debate (and most textbooks) is that *when firms exercise market power,* price caps do not have this undesirable side-effect. In fact, they can encourage firms to expand the amount they supply to the market (see the appendix for details). Given the mounting evidence that electricity generators are able to exercise some amount of market power, there are strong economic arguments in favor of judicious use of price caps in electricity markets. In fact, price caps are a common feature in deregulated electricity markets, although they are typically employed as a backstop measure leaving prices to be determined by the market under most circumstances.

Despite these theoretical arguments in their favor, price caps in the California market have had a checkered record, indicating that they must be designed carefully. At its inception, the California ISO implemented a $750 per MWh cap for electricity it purchased in the California real-time market. As the crisis heated up during the summer of 2000, the ISO lowered the cap twice to $250 in an effort to control prices. These caps did decrease the peak prices paid for power, but they also had negative side-effects. They tended to increase all bids submitted to the ISO, thus, the average costs did not decrease as much as expected. Second, when the prices of natural gas and NOx pollution permits increased, the cap was probably lower than the costs of generating power at many plants, deterring production. Finally, the ISO could cap the price of electricity bought only in California, creating incentives to export power to neighboring states (Bay Area Economic Forum, 2001a, p. 14).

At the peak of the crisis, the ISO completely abandoned its price cap in a desperate move to keep the lights on. FERC replaced those "hard" price caps with a "soft" cap of $150 in December, but wholesale prices soared above $300 per MWh.[3] FERC subsequently amended these with hard, regionwide caps in June 2001 at which time wholesale prices moderated substantially.

A number of lessons can be taken from this experience. First, regulators must be careful to not set the caps below the marginal cost of generation. Fortunately, the marginal production costs of electricity are well known because of years of cost and environmental regulations of power plants. Therefore, it is feasible to determine an appropriate level. This cap should be indexed to the costs of inputs such as natural gas and NOx permits to allow for changing market circumstances. Second, caps must be credible. If adjacent markets lack caps, generation will flee the state in search of higher prices. Also, when supply reaches emergency levels threatening blackouts, energy buyers may purchase energy "out-of-market" and pay prices above the official cap. This behavior creates incentives for players to engage in a disruptive game of chicken in which generators withhold supplies until the purchaser, desperate to avoid blackouts, is willing to pay above-cap prices. Effective price caps, thus, must apply to the entire region in which electricity is traded, not just one state in that region, and regulators must demonstrate discipline in upholding stated caps.

Third, price caps should be employed only as a temporary stopgap measure. If price caps are left in place over the long run, they may create disincentives to enter either the generation market or the retail market, thereby impeding transition to a workably competitive market in the long run. Also, experience from gas and oil markets indicates that enforcement becomes increasingly bureaucratic to deal with over time as producers demand exceptions and make efforts to evade caps (Hogan, 2001a). Finally, given that most of the financial damage inflicted by the crisis occurred as California and FERC debated over the appropriateness

---

[3]A "hard" cap means that no bid above that amount will be accepted. A "soft" cap accepts all bids below the capped amount. Bids above that amount are still accepted but must be cost-justified later.

of price caps, state and federal regulators must come to a clearer understanding of how and when price caps will be implemented *before* returning to a competitive environment.

Considering that at the height of the crisis thousands of megawatts of generating capacity remained out of operation, California policymakers must establish methods for ensuring that capacity is made available when it is needed. One approach that has been considered relies on direct inspection of facilities. It would mandate that the CPUC set maintenance schedules. If a plant does not operate when it is scheduled to be on-line, the CPUC would then inspect the plant to ensure that the plant is off-line for mechanical and not strategic reasons. This approach is limited, however, by the grave difficulties an outside party faces in evaluating plant operations. Generating plants are large, complex, and idiosyncratic operations. An inspector is in no position to second-guess the judgment of plant managers who have years of experience with, and knowledge about, specific operating needs. Alternatively, regulators could require that all plants supply power to the market except for days on which maintenance is scheduled and a limited number of days for unexpected problems. Beyond this allotment, for all days that a plant does not supply power, it would be responsible for acquiring an equivalent amount of electricity on the open market. In this way, regulators would shift the financial risk of mechanical breakdowns onto generators and avoid the unworkable task of inspecting plants.

Effective competition may also be hindered by the structure of California's generation market. The largest merchant generator, AES, controls only 4,700 MW of capacity, less than 10 percent of the market at peak summer demand. Applying standard measures of market concentration, a market composed of ten firms the size of AES would be deemed unconcentrated, implying that any single firm had little ability to influence market prices. There is increasing concern, though, that because of the real-time needs of electricity grids, similarly sized firms have much greater influence. On high demand days, even small generators may control the net margin of power—the difference between total load and power available from others. In those cases, that generator is in an unusually strong bargaining position because its power is absolutely necessary if blackouts are to be avoided.

In response to these concerns, the United Kingdom has forced the two largest privatized generating firms, PowerGen and National Power, to sell off over half their capacity (Bay Area Economic Forum, 2001b). FERC has also recently enacted new market structure rules that place much stricter constraints on the size of wholesale power producers (McNamara, 2001a). California policymakers must pay close attention to the size and market power of generators active in its market. Such vigilance would be especially warranted if PG&E succeeds in its efforts in bankruptcy court to transfer its significant generating assets to an unregulated entity. Having a larger number of firms, each controlling a smaller portion of the market, is likely to be desirable, although there is the danger of excessive divestiture preventing generators from taking advantage of economies of scale and scope, increasing overall industry costs.

Regulatory rules controlling the behavior of generators cannot alone prevent the exercise of market power. The market itself must operate effectively and discipline producers who bid high prices for the sale of their power. High bids are unprofitable when it causes generators to be left out of the market as the ISO matches demand with the lowest supply bids. These risks are greatest under two circumstances. The first is when supplies are ample, forcing multiple generators to compete aggressively to have their power dispatched. In California, the evidence shows that at low loads the markups for power are small but that they increase steadily as demand approaches system capacity (Bushnell and Saravia, 2002). Consequently, all the policies discussed in the previous section that maintain adequate reserve margins also help control market power by forcing generators to bid more aggressively. Controlling market power through high reserve levels is not cost free, however, in that rarely used capacity is expensive to build and maintain.

The second factor that can force more competitive bidding is increased demand responsiveness. If consumers curtail usage in response to price increases, generators who bid high are less certain that their power will be dispatched, forcing them to bid more aggressively. Moreover, demand responsiveness substantially decreases the benefits of withholding generation capacity unilaterally. If demand decreases, even slightly, with higher prices, the price increase generators can effect through withholding power decreases and the amount of power they can expect to sell at the higher price also decreases. The exact degree of demand responsiveness required to thwart unilateral withholding of capacity, however, remains uncertain, and even reasonably responsive demand may not suffice by itself to curtail the exercise of market power (Borenstein, 2001).

More detailed discussion of policy options for facilitating demand responsiveness are described below. The benefits of improved demand responsiveness extend beyond their effects on generator competitiveness. Thus, such policies should be pursued regardless of whether California continues with competitive markets or decides to emphasize public power or regulation. At this point, it is only necessary to point out that to the degree that California cannot rely on FERC to police market power abuses, market policies that maintain high levels of reserve and aggressively promote demand responsiveness will be necessary to protect consumers from market power.

Market Design

The many shortcomings in the design of California's electricity market must be addressed, although the exact amendments that should be made are a matter of debate. California can learn from other, more successful, electricity markets. Nevertheless, the design of competitive electricity markets remains an excruciatingly complex enterprise. All efforts at market restructuring have met with unexpected problems that have required midcourse corrections, raising doubts that a competitive, complete, and robust model already exists. A detailed discussion of market design is beyond the scope of this report. The recent market redesign proposal developed by the California ISO and submitted to FERC is over 200 pages long. Three central points, however, warrant mention.

The first is that the California market must allow long-term power contracts to have a much greater role. The ephemeral attractions of the spot market early in the restructuring process (e.g., increased regulatory transparency and low prices) camouflaged the risks and volatility that overreliance on spot markets entails. Greater use of long-term contracts will help control risks and improve the stability of the market. In addition, they can help mitigate the exercise of market power by expanding the markets in which generators must compete to sell their power. Competition for airline tickets, for example, does not occur only at the gate minutes before departure. Rather, competition is fiercest in the markets to purchase advanced tickets, when passengers are better able to compare prices and rearrange their travel plans. Passengers arriving at the last minute are typically forced to pay a premium price. The same logic applies to electricity markets, in that buyers can have a wider range of choices and more competitive prices when they purchase electricity in advance. The DWR has effectively accomplished this goal for the time being by locking in large quantities of power for as long as the next ten years. Nevertheless, regulators must still develop rules under which the utilities and other electricity service providers will enter into bilateral contracts, forward contracts, and long-term contracts.

It is important to note that although forward contracts increase competitiveness, reduce short-run price volatility, and mitigate market power, they do not, in themselves, guarantee lower prices for consumers. In the spring of 2000, for example, forward prices in California were less than $80 per MW and spot market prices hovered around $250. In contrast, at the same time New York forward prices were $140 and spot market prices turned out to be only about $80. Over the long run, forward prices in a competitive market will be similar to the average of spot market prices (Borenstein, 2001). In other words, although long-term contracts can protect consumers from price spikes, they will not protect them against higher prices resulting from persistently tight electricity supplies.

The second issue involves the degree of centralization of decisionmaking authority. California chose a design that relied much more on market transactions and less on centralized management by the ISO than other restructured markets did. This choice created a system that is especially unwieldy to manage during system alerts, as system operators scramble to maintain the grid. Policymakers should be concerned about such inefficiencies. There is strong evidence that vertically integrated utilities obtained significant operational savings from their central management of generator dispatch and transmission (Kwoka, 1996). Overreliance on markets for the dispatch of all specialized services required to maintain reliable grid operations risks squandering the benefits of centralized grid management.

The third issue concerning market design is that the way market rules are developed will be as important as the specific rules. The implementation of AB 1890 has been severely criticized. As one market observer described it:

> Ideological rhetoric played a bigger role than serious analysis or practical experience drawn from other countries. In the end, the ultimate design of the wholesale market institutions represented a series of compromises made by design committees including interest group representatives, drawing on bits and pieces of alternative models for market design, congestion management, transmission pricing, new generator interconnection rules, and locational market power mitigation. . . . Getting it done fast and in a way that pandered to the many interests involved became more important than getting it right (Joskow, 2001, p. 14).

The current environment still poses hurdles to a constructive debate. Many state policymakers mistrust the energy traders whom they blame for the energy crisis. Generators and energy traders are wary of the ISO as it has become politicized during the crisis, and California and FERC, who approves electricity market designs, remain on chilly terms. The ISO, who is responsible for market design amendments, operates in isolation from California policymaking bodies. Creating a forum in which these players and independent experts can come together is necessary to move the process forward.

Retail Competition

Retail competition was a major feature of California's original deregulatory framework, but its future remains in doubt, since the CPUC suspended direct access in September 2001. The impetus for suspending direct access was the need to repay the $42 billion worth of long-term contracts into which the state had entered. The price of the electricity purchased through these contracts has been significantly higher than spot market prices since the worst of the crisis passed. If consumers were allowed to turn to alternative suppliers, they could avoid paying for the more expensive state power, thereby increasing the costs borne by the remaining consumers.

Having all California consumers pay their fair share of the costs of the crisis is a worthy goal, but its pursuit need not derail efforts to promote retail competition. The state can retain policy flexibility by pooling the costs of the contracts that will exceed the future costs of power bought on the competitive market. The extent of these additional costs remains in flux because of pending cases before FERC and uncertainties over future wholesale rates and interest rates, but rough estimates range between $12 billion and $25 billion. California could then arrange to pay down this pool of excess costs through a nonavoidable charge to be levied on electricity users for the term of the contracts. A charge in the range of 0.7 cent to 1.5 cents per kWh would suffice. This charge would be similar in concept to the competitive transition charge that paid for the sunk costs faced by the utilities after restructuring. With such a charge in place, all consumers would help pay down these costs no matter which electricity provider they choose.

Reinstating retail competition would help promote a number of goals for the electricity sector. It would simplify regulatory tasks by reducing the need for oversight of the electricity sector. Even with a fully competitive wholesale market, insufficient competition at the retail level would leave electricity retailers free to charge monopoly prices to captive customers and would fail to provide incentives for retailers to control their electricity procurement costs. Consequently, it is only when consumers are able to compare and choose between a variety of providers that the market forces can replace regulation as the method of disciplining electricity service providers.[4]

Retail competition can also play an important role in controlling market power. The opportunity to choose between providers can stimulate individuals and firms to become more active and intelligent consumers, and if this stimulates demand-side responsiveness, it will constrain the ability of generators to demand high prices for their power.

---

[4]Even with competitive electricity service providers, regulation will still be necessary to ensure the financial stability of ESPs. This oversight, however, is more like the regulation of financial institutions, such as Savings and Loans, than traditional utility price regulation. See Wolak (2001) for details.

Finally, competition between service providers can also stimulate the creation of innovative price and service options, leading to increased efficiency in electricity production and consumption. We could see an effect similar to that in the cellular phone industry, which offers a much wider range of pricing and service plans than the regulated wireline telephony does. Such potential innovations include real-time pricing options, long-term contracts for end users, and energy-management services.

The extent of these benefits and the time frame in which they are realized, however, depend on the strength of retail competition, the innovativeness of retailers, and the active involvement of consumers. Despite its attractiveness, retail competition has developed slowly and has suffered setbacks in other markets. Firms specializing in retail electricity service have not been performing well, and even in states considered a success story in terms of retail competition, such as Pennsylvania, the vast majority of customers remain with incumbent utilities (McNamara, 2001b). There are instances in which ESPs have offered innovative price and service packages, but these have, for the most part, targeted large customers. Efforts to lure small customers, in contrast, have focused on simple rate discounts or green energy. This early record is not surprising given that competition often takes time to develop in formerly regulated industries. It does raise questions about the degree to which residential and small business consumers are willing and able to analyze their electricity purchases actively, but as experienced in other deregulated markets, such as telephone equipment and long-distance calls, consumers have become accustomed over time to making new choices among multiple providers.

Regulators face a number of choices concerning the implementation of retail competition. The right mix of policies can promote the speed with which competition develops and improve the efficiency in the retail market, although these choices generally come at the expense of increasing the volatility of consumer bills and the complexity of consumer choices. The first choice is the default provider—the firm assigned to provide electricity to consumers who do not actively select a provider. Assigning this role to the incumbent utilities simplifies life for passive consumers who continue to be served by the same firm but increases the barriers to entry for new ESPs. Alternatively, the default service can be auctioned off to the provider or providers that offer the lowest price offering. Such an auction helps establish the correct price for default service and facilitates entry. Another policy that can promote the development of retail competition is to assign customers randomly to qualified ESPs. This system allows new entrants to get established quickly, although it forces some customers to switch providers against their will.

The second choice relates to the level and structure of the default offering. The price can be either fixed or vary with underlying wholesale rates. Fixed rates offer consumers added stability, although they must be tied to the long-run underlying costs of power (e.g., either long-term contracts for power or costs of owned generation) if California is to avoid repeating the financial crisis from which it is only now emerging. Variable default rates have the benefits of improving efficiency and forcing consumers to be more active, but they place greater risk and complexity on consumers. Variable rates are discussed in greater detail in the next chapter in a discussion of demand responsiveness programs.

The level of default rates is another important variable. Setting the rate low protects consumers, but it makes it difficult for new service providers to enter the market profitably. A higher default price gives customers greater incentives to experiment with new electricity service providers and facilitates new competitive entry. Higher default prices, however, disadvantage passive consumers.

Transition Strategy

Developing a well-functioning electricity market in California also requires a transition strategy. The success of other restructuring efforts demonstrates that markets can be made to work. Nevertheless, they are complex, and successful markets have numerous components, all of which must be operating for the system as a whole to function. These components include, among other things, a sufficient number of competitive wholesale generators to yield competitive results, active spot and forward markets enabling market participants to hedge risks, a competitive retail market that offers a wide range of price and service options to accommodate consumers' differing risk preferences and consumption habits, and consumers who are able and willing to manage their electricity consumption intelligently. Given that institutions and behaviors develop only slowly over time, not all of these components can be in place immediately. The transition stage—when some components, such as wholesale markets, are in place whereas others, such as retail competition and hedging of market risks, remain underdeveloped—poses significant risks of which policymakers must be wary.

A successful transition requires sensitivity to the interdependencies among these components and attention to getting them all in place. Actions must be carefully ordered. For example, active retail competition must be established before lifting retail price controls if consumers are to be protected during the transition. Similarly, a more controlled transition toward retail competition may be warranted. Large users, who have greater capacity to manage their consumption, could be offered competitive choices first, with residential and small business users being given more time to adjust. Finally, a transition strategy must include mechanisms for addressing unexpected problems as they arise and facilitating midcourse corrections.

Hybrids

Policymakers have expressed interest in hybrid industry structures that entail various combinations of government ownership, rate-of-return regulation, and competition. These include continued wholesale competition in conjunction with regulated retail markets, competitive markets with active participation of government-owned entities, and schemes that differentiate among consumer groups, allowing some, such as large industrial and commercial users, to shop for competitive power while continuing to provide regulated power for other groups, such as residential and small business users. The attraction of these hybrids is that they appear to enable policymakers to pursue what are otherwise conflicting goals and they provide a smoother transition path toward a final industry structure.

Such hybrids, nevertheless, must be designed carefully. The roots of the California crisis can, in part, be traced to the pursuit of multiple goals, each of which was valuable and reasonable in isolation. Developing wholesale competition for power made sense as a means to enhance efficiency. The retail price freeze enabled the utilities to recoup their stranded costs. The 10 percent rate reduction provided to residential and small business customers could be supported as a means to ensure that the benefits of competition were shared by all customer classes, and the constraints the CPUC placed on long-term contracting were needed as a means to protect captured consumers from inflated contract prices and to promote regulatory transparency.

In combination, however, the pursuit of these goals created an explosive mix. The rate reduction and freeze stifled retail competition, because small customers had little incentive to explore alternative service options. The lack of retail competition, in turn, bolstered the CPUC's resolve to review the prudence of long-term contracts. Both of these then increased the utilities' exposure to the spot market, which led to financial disaster when spot market prices shot up.

The interactions of policies that pursue different goals are therefore paramount, and policy designs must combine elements that are compatible. One principle that should be followed is that utilities and other energy service providers must balance the term structure of their retail and wholesale transactions. If an ESP buys power on the spot market it must sell power at prices that fluctuate with the wholesale spot market. In contrast, if an ESP provides its customers with fixed prices for a specified length of time, those sales should be backed up by fixed-price, long-term contracts of similar duration. With such a balance, the retail price freeze implemented in AB 1890 was an achievable policy goal if the utilities had entered into long-term contracts. Conversely, forcing the utilities to buy on the spot market was also a legitimate goal, but only as long as consumers paid rates based on spot market prices. It was the combination of a price freeze with spot market purchases that led to disaster.

Another principle is that the roles of competition and of price and quality regulations must be carefully balanced. They rarely coexist in the same market successfully over the long run. Continued regulation can thwart the emergence of competition by impeding new entry and customer choice. At the same time, sufficient competition must exist for consumers to benefit from deregulation. This can lead to a vicious circle in which neither effective competition nor coherent regulation prevails.

If insufficient competition exists to justify immediate deregulation, regulators must decide whether to maintain existing regulations or to develop a clear path that promotes entry and active consumer choice so that competition can prevail. In addition, mixed regulatory schemes in other industries, such as telecommunications, have typically led to artificial distinctions between service and customer categories. These distinctions create a host of opportunities for market participants to game the system, complicating regulatory tasks and leading to unintended consequences.

The strict tradeoffs between policies that promote stability and those that promote efficiency are also underappreciated. Improvements in system efficiency depend on firms and individuals being able and willing to respond to incentives by changing, among other things, the way they run plants and how they consume electricity. At the same time, firms and individuals seek to buffer themselves from environmental vicissitudes and risks because it is expensive and difficult to be constantly changing one's routines. Such buffers, however, reduce the incentives to engage in efficiency-enhancing actions. Consequently, improved stability comes at the price of reduced efficiency, yet there is continued confusion about this tradeoff. It is not widely understood that long-term contracts are likely to include a premium for the price stability they provide the purchaser. More generally, the tradeoff is often treated unevenly. The same analysts who advocate exposing consumers to greater price volatility also advocate that utilities should sign more long-term contracts to mitigate the volatility of spot market energy purchases. If utilities benefit from long-run price stability, it is certainly at least as valuable for consumers.

Conclusions

Table 4.2 presents a summary of how the main institutional alternatives satisfy the main goals for the electricity sector. Overall, policymakers face a choice between the greater stability, reliability, and administrative feasibility of regulated utilities or public ownership versus the prospects for greater efficiency gains through competitive markets. In terms of environmental protections, no market regime clearly

Table 4.2

Alternative-Criterion Matrix: Institutional Alternatives for California's Electricity Sector

	Public Ownership	Reregulate	Market	Hybrid
Low prices	Good historical record Able to target favored groups Historical advantages unlikely to be available in future	Good historical record Able to target favored groups Failures in 1980s led to deregulatory push	Promoting competition requires higher default rates Competition can drive prices down; market power is a concern Results from recently deregulated experiments uncertain	Can target low prices to certain consumer groups
Bill stability	Weaker record in California than in regulated monopolies	Strong record	More volatile prices Can be mitigated with long-term and hedge contracts	Can protect certain consumer groups from volatile wholesale rates
Efficient resource use	Similar to regulated private monopolies	Weak incentives Bias toward excess capacity Can be improved with marginal cost pricing and incentive regulation	Potential for significant improvements Inefficiencies resulting from vertical disintegration require careful design of markets	Hybrids restrict efficiency gains from competition Can create perverse incentives
System reliability	Good in long run Risks of transition	Strong historical record Long tradition of attracting investment Future investment in post-crisis environment less certain	Market-based investment susceptible to boom-bust cycles Poor coordination at times of system crisis Requires much greater demand-side management	Mix of design features must be chosen carefully

Table 4.2 (continued)

	Public Ownership	Reregulate	Market	Hybrid
Administrative feasibility	No state or local structures exist to run major power effort Will face concerted political resistance	Rules are transparent and well understood Buy-backs of generating capacity may not be feasible Regulating procurement with long-term contracts is feasible Likely resistance from utilities	Complex market design and regulatory issues Regulatory control shared with FERC Requires active consumer decisionmaking	Systems of mixed rules are complex and contentious
Environmental protections	Public managers retain discretion to pursue environmental goals	Success with demand-side management Bias toward overcapacity leads to environmental damage	Could be degraded if environmental review is slackened Created market for green power Competitive firms may respond better to environmental incentives Improved efficiency decreases environmental damage	
Other	Less desirable in generation market as it becomes structurally competitive More appropriate for transmission and distribution	Less desirable in generation market as it becomes structurally competitive Still necessary for transmission and distribution monopolies		

dominates the others, mainly because environmental results depend on complex interactions between each regime and existing environmental regulations.

The nature of the tradeoffs differs, however, for different segments of the electricity industry: generation, transmission, distribution, and retail marketing. The arguments in favor of competitive markets are strongest for generation. This segment of the industry has been most affected by technological changes that have led to ever-smaller plants that generate power at competitive costs. State-of-the-art dual cycle gas-generating plants have a capacity of about 500 MW, a fraction of the size of California's market, and cogeneration facilities are even smaller. These plants have enabled new players to enter the market as power producers and large power users to self-generate economically. Consequently, the California electricity generation market is wide open with hundreds of private and public entities owning plants. In such a world, regulation or public ownership becomes increasingly anachronistic. As long as a competitive environment can be maintained, reliance on multiple providers each competing against the others is more likely to provide reasonable service than depending on the good performance of a single monopolist.

In the short run, policymakers may choose to restrain the development of competitive generation markets if they wish to promote a more stable electricity sector and are wary about ceding control to FERC for mitigating the market power of competitive generators. Nevertheless, they should be aware that the march of technology will continue, making it increasingly difficult and inefficient to bottle up alternative providers. Planning for an eventual transition to a more competitive market is important, and policymakers need to avoid choices that will impede such a move in the future. Specifically, they should avoid forcing the regulated monopolies to buy or build additional capacity in the short run. Such policies simply increase their market power, impeding the development of a competitive market in the future. Similarly, it would be a mistake to assign the full costs of the state's long-term contracts to the utilities with the expectation that regulated rates will allow them to recover these costs fully. Such a move would create a significant cost difference between utility and merchant generators, leading to political impediments to opening up their markets to competition.

Transmission and distribution functions remain monopolies, making some form of regulation necessary to ensure reasonable rates and open access to all generators and final users. Public ownership is a feasible alternative, and the state's aborted attempt to purchase the utilities' transmission lines in exchange for financial bailouts would have made sense at the right price. Other attempts at public buyouts of transmission and distribution assets, however, will be equally controversial and expensive. Continued regulation of privately owned assets offers the most reasonable and well-understood alternative.

In the retail segment, the tradeoffs between regulated versus competitive structures depend on consumers. Potential efficiency gains from competition are derived by changing consumer behavior, making them more aware of the real costs of electricity and compelling them to change their consumption accordingly. These gains can come about, however, only if consumers are fully exposed to price volatility and are willing and able to manage that volatility.

If consumers wish to be shielded from such volatility and wish to remain passive consumers of energy, the benefits of a competitive regime are reduced. Resistance to price hikes and more complex pricing proposals suggest that consumers are not interested in being exposed to price volatility. Opinions, though, probably differ between different customer groups (e.g., residential and small business versus larger business users) and can change over time as customers understand how competition can enable them to reduce their overall energy costs. These differences suggest that a hybrid model with retail competition introduced in stages, first to larger customers and only later to smaller customers, offers important advantages.

5. Overarching Recommendations

While California policymakers grapple with the fundamental issues of reconstituting the market and regulatory institutions of the electricity sector, they should take a number of actions to make it more robust no matter what reform path is taken. The California crisis exposed a number of weaknesses in the management of the sector. Early in the crisis, California lacked mechanisms to elicit consumer conservation in response to tightening electricity supplies and rising wholesale prices. The absence of a demand response exacerbated the crisis before the summer of 2001 when considerable conservation efforts were made, helping to tame wholesale rates. The crisis also revealed the interdependencies among components of the state's energy infrastructure and the dangers of neglecting any single component. Inadequate transmission capacity, for example, exacerbated shortages in generation capacity, and a heavy reliance on gas-fired generation made California particularly susceptible to disruptions in the natural gas market. Finally, the crisis highlighted the need for effective and responsive policymaking to manage this increasingly complex and volatile area.

The successful development of an efficient, low-priced, and reliable electricity system depends on addressing these weaknesses. This chapter discusses actions the state can take to improve demand responsiveness, develop a more comprehensive infrastructure strategy, and reorganize and clarify its policymaking functions.

Improve Demand Management

The implementation of AB 1890 focused almost entirely on the supply side of the electricity market, working to create a competitive wholesale power market. Reforms of the demand side of the market were, in contrast, ignored and often undermined. Regulators failed to promote retail competition. Funding for conservation programs was reduced, and consumers were shielded from price fluctuations. As policymakers continue to seek ways to lower the costs of electricity and improve the efficiency and reliability of the system, demand-management policies cannot be left out of the equation.

Successful conservation efforts contributed significantly to the unexpected passing of the crisis at the beginning of the summer of 2001. Reductions in peak demand amounting to as much as 14 percent in July helped avert the widespread blackouts that had been predicted. It is important to note that these extraordinary conservation efforts were achieved without derailing the California economy or causing undue individual hardships. In fact, in a survey of Southern California Edison customers conducted for the California Energy Commission, 70 percent of respondents said that their conservation efforts had either no serious effect on their lifestyle or even possibly improved their lifestyle (California Energy Commission, 2002a).

Policymakers implemented the conservation measures in 2001 as a crisis response, but they should not think of these programs as solely emergency measures. Conservation programs offer significant potential benefits under a wide range of market conditions and regulatory environments. Reductions in peak demand, for example, can decrease the costs of generating electricity by reducing the need for investments in peaking capacity and transmission plant, thereby reducing prices. California has a long and successful record of incentive programs promoting investments in energy efficiency, and these efforts have been found cost-effective in comparison to investments in additional generating capacity. For individual users, the rationale for increasing energy efficiency has never been stronger. Because Californians will face significantly higher rates for several years because of the crisis, the savings from conservation are especially great.[1]

---

[1]It is true that high electricity prices give individuals and firms greater incentives for conservation, but the same is not true for the state as a whole. Higher electricity rates going forward will primarily pay for costs that have already been incurred, the cost of the long-term contracts signed during the crisis, and debt accrued by the utilities. These sunk costs cannot be avoided even if the state dramatically and permanently reduces its overall demand. Future conservation efforts by the state save only the avoided costs of additional

Demand-side management can also play an important role in improving system reliability. Because energy consumption can be altered more quickly than new generation can be brought on-line, demand responsiveness is important for keeping supply and demand in balance, especially in times of tight supply. Finally, such programs are also environmentally friendly, by reducing emissions and avoiding the construction of additional plants. These programs should be a standard feature of a well-functioning electricity sector and need to be expanded and made permanent.

Traditionally, demand-management programs have focused on conservation and enhancing energy efficiency. One set of programs has focused on developing and imposing efficiency standards for building construction and appliances. A significant improvement in the efficiency of air conditioners, for example, could forestall the need for additional plants because air conditioning represents as much as two-thirds of usage during summer peak hours. Other programs have been implemented to identify opportunities for efficiency-enhancing investments and to create incentives to make those investments. For example, since restructuring, the CPUC and CEC have administered a public goods program that has collected a fee from ratepayers and allocated grants for efficiency-enhancing investments. The recently formed California Consumer Power and Conservation Financing Authority has also been given the power to invest up to $1 billion in conservation programs, although these programs remain under development.

Another method is to offer direct incentives for conservation. An example is the 20/20 program that was a centerpiece of the state's conservation efforts in 2001. It offered consumers a 20 percent rebate if they reduced consumption by 20 percent from the previous year's levels, and over 30 percent of utility customers qualified for the discounts. Finally, education and outreach programs can heighten consumer awareness of their electricity usage and disseminate information on easy methods for reducing consumption, such as shutting off a little-used spare refrigerator. Although it is difficult to disaggregate the individual

---

power, and with wholesale markets working more competitively, these prices are significantly lower than those reflected in current rates.

effects of the many conservation programs that were implemented concurrently, public information programs do appear to have played an important role during the crisis.

Such programs are useful because residential and business consumers are often constrained in their ability to research conservation opportunities, calculate their costs and benefits, and raise the capital for investments. Consequently, standards and well-designed incentive programs can play an important role in aiding consumers with this investment decision. Standards are also useful in that they are administratively easy to implement and can lead to relatively rapid changes in consumption patterns.

These programs, however, do have limitations. Successful programs are difficult to design. The effects of standards depend on the way technologies are used and the future prices of electricity. A major risk with standards is that they can be set too strict, at which point the added costs of efficiency enhancement outweigh the benefits of lower energy bills and reduced need for capacity expansion. Incentive programs can also be inefficiently expensive. Such programs risk rewarding consumers for actions they were already intending to make, such as buying an energy-efficient air conditioner or voluntarily reducing consumption. The 20/20 program under certain circumstances could result in payments far exceeding the costs of the saved power.[2] Such programs also miss conservation opportunities because regulators cannot identify every possible savings within the idiosyncratic energy-consumption patterns of residents and businesses. Most important, although these energy-efficiency programs reduce overall usage levels, they do little to change usage in response to market conditions such as at times when supplies are tight.

---

[2] A consumer who used 1,000 kWh during one month last year and received a $100 energy bill would have to reduce consumption to 800 kWh for the same month to receive a $20 rebate under the 20/20 program. With a voluntary reduction in consumption to 850 kWh in response to a public information campaign, participation in the 20/20 program would yield only an additional reduction of 50 kWh. In this case, the costs of added conservation are $0.40 per kWh (a $20 refund for a 50 kWh reduction). These costs are far above the average cost of wholesale power even during the worst months of the crisis.

These traditional demand-management programs can be made more effective and more pervasive by expanding them to induce consumers to make consumption decisions that respond to underlying market conditions. The key to these changes is to give consumers information on the cost of power. The cost of generating power fluctuates constantly throughout the year and each day. Figure 5.1 provides an example of these fluctuations over one summer and one winter day.[3] These hourly and seasonal price variations are almost never passed through to consumers. Rather, they usually receive a constant average price over all hours of the day throughout the year.

Passing through these price signals to consumers offers a number of advantages over traditional conservation programs. Because power prices are highest when demand is high and supplies are short, consumers face strong financial incentives to control their usage when conservation is

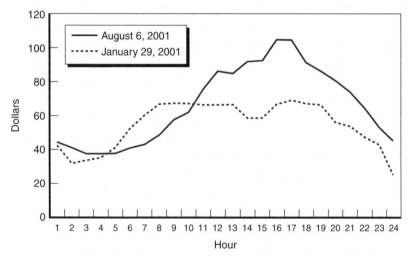

SOURCE: http://www.mis.nyiso.com/public/P-2Alist.html.

Figure 5.1—Day-Ahead Wholesale Market Price Hourly Fluctuations for New York City

---

[3]These wholesale market prices indicate only the underlying marginal costs if power markets are fully competitive. If generators are exercising market power, these wholesale prices may overstate actual costs, but the prices in either case do correctly reflect the pattern of cost changes over time.

most valuable. Price averaging, in contrast, impedes efforts to induce consumers to manage their usage because their savings from curtailing usage in the middle of the day (when conservation is most needed) are no greater than their savings from curtailing usage in the middle of the night. The 20/20 rebate program had the same limitation in that it rewarded demand reductions during nonpeak hours as much as on-peak reductions. Unlike grants for efficiency-enhancing investments or rebates for conservation, price signals also automatically adjust to changing market conditions. Thus, if shortages develop suddenly and unexpectedly, as they did in the summer of 2000, conservation measures are put into place immediately.

The most market-oriented method to create incentives for conservation at times of peak loads is to charge consumers real-time prices that change every hour with fluctuations in the wholesale price. RTP continuously provides consumers with the correct information on the value of conservation and the incentives to carry it out. At this time, however, RTP is most appropriate for industrial and large commercial customers. They have the capacity to manage their electricity usage in real time and to hedge risks that they face as a result of price volatility. RTP also requires more sophisticated metering technology that records the time of usage. Such meters are not commonly installed at customer premises, although in 2001 California made an important step toward implementing RTP by installing 23,000 real-time meters to large customers that consume 25 percent of the peak load.

Residential and smaller businesses, in contrast, are less likely to manage their electricity consumption effectively in response to continuously changing prices. Thus, simple RTP programs are less likely to induce reasonable conservation efforts. Nevertheless, several variations of RTP programs greatly simplify the decisionmaking process, allowing smaller consumers to respond to prices more effectively. One option is interruptible rates. This program offers a discount to users who are willing to have their power curtailed occasionally when power is particularly scarce (and prices are particularly high). These programs were popular before the crisis. They fell out of favor when the number of power interruptions grew dramatically, but they continue to be a cost-effective way to manage demand. Another method is time-of-use (TOU) rates under which power consumed at peak times is priced at a higher level than off-peak power. Although TOU rates provide some incentive to conserve at the correct times of the day, they still fail to reflect changing day-to-day circumstances. For example, consumers would face the same peak hour TOU rate on each of the two days represented in Figure 5.1 even though the cost of generating power on the summer day is significantly higher.

The effectiveness of TOU rates can be improved by a system called critical peak pricing. In addition to the higher on-peak rates, critical peak pricing raises prices to very high levels a few days each year when the system approaches its capacity. Warnings that these higher rates will be in effect can be broadcasted the day before they occur, allowing consumers time to plan a response. One simple option during the summer would be to precool one's house in the early afternoon, before the peak prices take effect, and then turn off the air conditioner for the late afternoon hours.

One promising avenue for helping small customers to respond effectively to market prices combines real-time meters, automatic thermostat technologies, and other electronic control technologies. These technologies enable electricity service providers and consumers to automate responses to price fluctuations, thereby greatly simplifying demand management. For example, one system would offer customers a reduction in the standard tariff. In exchange, they would grant their service provider some control over their air conditioning use. On a set number of days, the provider could curtail the customer's electricity use by increasing the temperature setting for the air conditioner. It is also possible to offer customers an override feature. The customer could choose to lower the temperature but would then be charged the going market rate for power. In another situation, customers could program a set of decision rules into their thermostats: If the price of electricity is less than $0.10 per kWh, they may wish to cool their house to 72 degrees. If the price exceeds $0.10, the house could be allowed to warm to 76 degrees, and if the price exceeds $0.25, the air conditioning could be automatically shut off. As the costs of these technologies decrease, the possibilities expand and virtually all household appliances could be controlled automatically according to the price of electricity.

Experiments are already being conducted with a number of these options. Much will be learned in the near future on customer acceptance of these programs and the degree to which they facilitate intelligent energy management.

Although the potential benefits of RTP and similar programs are significant in terms of improving system reliability, reducing costs, and improving efficiency, they have met with strong resistance. Consumers find these programs complex and uncertain and fear that they will result in higher overall bills. Some of these fears are based on misconceptions. Businesses that have continuous operations, requiring a constant amount of power all hours of the day, have complained that they should not be exposed to real-time prices because they cannot avoid using electricity during the expensive afternoon hours. Such businesses, however, would benefit from RTP because a smaller proportion of their total usage is during peak hours compared to the average user. Thus, the savings from the large quantities of low-priced electricity they purchase at off-peak times would outweigh the added expenses for afternoon operations.

Nevertheless, some concerns about exposing consumers to the full variability of electricity costs are real. They can lead to increased volatility in monthly bills because spot market prices vary considerably from month to month. Moreover, because consumers can take only a limited number of actions to reduce their electricity consumption, energy demand does not change significantly in response to price changes. In the summer of 2000, for example, SDG&E customers faced a 140 percent price increase, but they decreased usage by only 5 percent. Politicians may have diminished the price response by mobilizing to reregulate rates, thereby convincing consumers that the higher prices would not endure (Bushnell and Mansur, 2001). Nevertheless, studies generally find that demand responsiveness is limited. Large price swings, consequently, are required to induce changes in behavior.

This volatility can be mitigated. Automatic thermostats and control technology help by automatically reducing usage when prices are particularly high. Customers could be allowed to purchase forward and hedge contracts that could insure them from the effects of extreme price swings (Wolak, 2001). Within a regime with direct access, competitive service providers could offer a variety of such contracts, although the decisionmaking calculus for many consumers could prove overwhelming.

Even in the absence of retail competition, utilities could be mandated to offer RTP while maintaining bill stability by requiring that they hedge a large portion (e.g., 80 percent) of their retail load with fixed-price, long-term contracts. Customers would then be charged the hourly RTP minus the difference between the average real-time price and the utility's actual procurement costs (e.g., the costs of the long-term contracts plus the additional power purchased on the spot market) (Borenstein, 2001). For example, if the average real-time wholesale market price is $0.15 in one month, long-term power costs $0.10, and the utility purchased 20 percent of its power in the spot market, then the customer is charged the RTP minus $0.04 for each kWh.[4] This method preserves the incentives to conserve power at system peaks because the hourly rate paid by the consumer tracks the real-time prices closely, but it also maintains average monthly bills that are close to the costs of long-term power.

A final option implemented by Georgia Power is to provide customers with a historical load profile that estimates their power usage for each hour of the day. Customers are then charged real-time prices only for deviations from this historical load profile. Customers who do not change usage patterns at all receive exactly the same bill as they did in previous years, but if they conserve (use) additional electricity at peak times, their bills are reduced (increased) by the prevailing real-time price.

Certain types of users do end up with higher bills under RTP and similar options. Consumers who consume a disproportionately large amount of energy during peak times and are unable to shift their usage patterns get stuck paying the higher peak prices. Such customers would include, for example, firms that do most of their business in the afternoon and residences that rely heavily on air conditioning, but for health or other reasons cannot curtail use at peak times. The Georgia Power policy of charging real-time prices based on deviations from historical usage patterns is one method for mitigating these harms. Education programs to help consumers understand ways to shift their

---

[4]The average cost of procurement in this example is $0.11 = 0.8 × 010 + 0.2 × 0.15.

usage away from peak times would also reduce the scope of the problem. For the small number of users that are completely unable to change their usage patterns, targeted subsidy programs could be considered.

Develop Comprehensive Infrastructure Planning

As California restructures the electricity sector, an important focus will be strengthening the state's capacity for comprehensive infrastructure planning. Although the crisis was driven by a shortage in electricity-generating capacity, it also revealed weaknesses with the transmission system, natural gas supply, and other related systems. Simply building additional generation plants will not solve the state's long-run problems if inadequacies with complementary systems are not addressed.

Restructuring and crisis have led to chaotic systems for monitoring, siting, and building additional transmission capacity and gas pipelines. Before restructuring, responsibilities were clearly delineated. The main utilities overseen by the CPUC built and maintained the main transmission grid, and the CEC approved connections to the grid from generating plants. With restructuring, the responsibility for improvements for the transmission system was dispersed among many agencies, allowing transmission planning to fall through the cracks. At the same time, restructuring demanded more not less attention to the transmission grid. A boom in energy trading created new flows of power over the grid that stressed a system that had been designed by vertically integrated monopolies at a time when power flows were more predictable and controlled.

Developing plans and implementing improvements to the grid is certain to be contentious and difficult. Electricity transmission remains a monopoly and, thus, it is a continuing regulatory challenge. In the ideal vision of electricity market design, wholesale market trading was to create incentives for transmission expansion. Firms would have to pay for congestion on the system, and these congestion prices would signal the need for new transmission capacity. The California market has clearly failed to provide these incentives and instead created opportunities for gaming the market (Hogan, 2001b).

Expanding transmission capacity also creates conflict by shifting the competitive landscape for generators. Generators within zones served by inadequate transmission links tend to receive higher prices because congestion prevents generators outside that zone from importing competitive power. Expanding capacity to such zones reduces the prices generators within the zone can charge while expanding the market for generators outside of the zone. Given the competitive stakes, transmission siting decisions can lead to protracted regulatory and legal battles. Finally, new transmission facilities face environmental hurdles because they are unsightly, and there is continuing debate on whether exposure to electromagnetic fields created by high-tension wires can cause cancer (Oak Ridge National Laboratory, 1998). To overcome these hurdles, California requires a comprehensive, coordinated, and committed transmission policy.

The state should also support research and development in methods for making California energy infrastructure more flexible and robust. Renewable forms of energy are attractive because of their environmental benefits, but they also should be examined as a method for improving the diversity of fuels on which California relies. Innovations in new forms of generation also hold promise as a way to alleviate future supply shortages and bottlenecks (Cambridge Energy Research Associates, 2001a; Cicchetti, 2001). Microturbines and fuel cells are two technologies that are beginning to be available or are in development. They are sufficiently small and efficient that large customers could become responsible for their own power. Such distributed generation can be added quickly to relieve shortages from bottlenecks in the transmission grid. In addition, micro sources of generation allow for mobile generation in which power can be moved to areas in need. By reducing the level of reserves needed to maintain system reliability and by using available transmission resources more effectively, distributed generation could enhance the efficiency of the electricity system.

Reorganize Policy Apparatus

Whatever direction the development of the California electricity sector takes, policymakers must reassess and reorganize the complex set of administrative structures that currently exist. Electricity sector restructuring followed by crisis has led to an ad hoc and confusing mix of state agencies and departments. Before AB 1890, California, the CPUC, and the CEC shared primary responsibility for managing electricity policy. They regulated electricity service and rates, approved new plants, performed long-run planning, and ran energy conservation programs. The development of a competitive market for wholesale power ceded some regulatory authority to FERC. It also led to the creation of the ISO, an independent market body regulated by FERC, and a new state agency, the EOB. In the heat of the crisis, the Department of Water Resources became the default purchaser of electricity for the state, and the legislature create the California Power Authority in an attempt to regain some control over the situation.

As seen in Table 5.1, these changes have led to overlapping, confused, and conflicting authority. The California Power Authority now has planning and conservation responsibilities similar to those of the CEC and CPUC. The CPUC, CEC, DWR, and ISO all forecast energy demand and supply. Responsibility for monitoring electricity system reliability is split between the CPUC, EOB, and ISO, and the CPUC,

Table 5.1

Selected Activities and Responsibilities of Energy-Related Agencies

Function	CPUC	CEC	CPA	DWR	ISO	EOB	FERC
Rate regulating	X						X
Promoting energy conservation and efficiency	X	X	X				
Forecasting electricity demand	X	X		X	X		
Promoting renewable resources		X	X	X			
Licensing generators		X					
Conducting integrated resource planning	X	X	X				
Monitoring the electricity market	X					X	X
Monitoring/planning system reliability	X				X	X	
Planning electricity transmission infrastructure	X	X	X		X	X	X
Planning natural gas infrastructure	X	X	X				X
Representing the state at FERC	X			X		X	

SOURCE: Adapted from California Legislative Analyst's Office (2002).

EOB, and DWR all share the duties of managing California relations with federal regulators.

These overlaps have led to coordination and policy failures. Most notably, the division of ratemaking authority between the CPUC and FERC impeded rapid and effective policy action at the height of the crisis. After the DWR entered into $42 billion worth of long-term power contracts, it clashed with the CPUC. The DWR demanded authority over rates to ensure that it could cover the contract costs and the bond issue to repay the state general fund for earlier electricity purchases. The CPUC, who traditionally set rates, refused to cede its authority, delaying the bond issue and costing taxpayers millions in added interest charges. The fractured responsibility over transmission projects has led to interjurisdictional turf battles and delayed much needed expansion to expand Path 15. In the wholesale power market, the new mix of state activities has given rise to potential conflicts of interest. The state through the Department of Water Resources and potentially through the California Power Authority competes with merchant generators in the wholesale market, but at the same time, it regulates these firms through other state agencies and the state's control of the ISO board.

This ad hoc structure of California energy policymaking institutions is an impediment to attaining the basic goals of the electricity sector. Administrative feasibility is hampered by the need for interagency coordination and policy implementation is impaired by confused program authority. State energy policy loses its coherence as the many, interrelated facets of energy policy—regulation of market players, market design, siting of generation and transmission, conservation, planning— are addressed in separate forums. Moreover, administrative conflict and chaos threaten the reliability and efficiency of the electricity system. Power generators may steer clear of constructing additional capacity in California if they are uncertain of the rules and regulations that will determine the returns on their investment. Similarly, consumers can become quickly confused if presented with a patchwork of differing options promoting conservation efforts.

Either an umbrella organization, such as a cabinet-level post, is required to coordinate policy or functions need to be rationalized and centralized into fewer key agencies. The exact shape of the necessary reforms depends on the institutional course that California follows. If the state continues to manage its portfolio of long-term contracts and expands the public role in the power sector, these functions could naturally be organized within an expanded California Power Authority. If policymakers wish to return to a more regulated environment, the CPUC would be the natural agency in which to centralize functions. In contrast, if the state wishes to forge ahead with a private, deregulated market, a new agency or a stripped-down version of an existing one could focus on a more limited set of regulatory functions, such as plant approval, conservation promotion, and market oversight.

6. Conclusions

The brunt of the electricity crisis has passed. The direst predictions for the "perfect storm" of the summer of 2001 did not come to pass, and recent additions to generating capacity appear to provide adequate supplies for the near future.

The California electricity sector, nevertheless, remains in serious condition. California businesses and consumers will be burdened with the costs of the crisis for years to come, placing a drag on the state's economy. The medium-term supply situation is also tenuous. Falling wholesale power prices combined with the fallout from the Enron bankruptcy have dried up the supply of investment capital available to fund power projects. As a result, merchant generators have cancelled or delayed thousands of megawatts of planned construction. These cancellations leave the state vulnerable to future supply shortfalls if there is a repeat of extreme weather conditions or demand grows rapidly because of slackening conservation efforts or unexpectedly strong economic growth. The most serious point is that California electricity policy remains adrift, lacking a long-run vision of how to move beyond the debacle of 2001.

The crisis did provide policymakers with some hard-earned lessons:

- **Electricity market design is fraught with difficulties.** The high costs of reliability failures and the need to balance supply and demand in real time greatly complicate the coordination of multiple players through bidding processes. Although there is no broad consensus on the optimal electricity market design, California clearly had significant deficiencies.
- **Heavy reliance on wholesale spot markets is extraordinarily risky.** As in other commodity markets, spot prices are very sensitive to shifts in underlying supply and demand conditions, at times leading to extreme volatility. In times of tight supply,

spot markets work particularly poorly, giving generators opportunities to flex their market power.

- **Fragmented regulatory authority impedes effective policymaking.** The division of regulatory authority between FERC, which controls wholesale rates, and the CPUC, which controls retail rates, was a recipe for confusion, blame shifting, and eventually disaster. More generally, the multiplication of electricity-related agencies led to duplication, confusion, and conflict in policymaking.

- **Competitive markets require behavioral adjustments.** During the implementation of AB 1890, regulators, utilities, and consumers were all slow to recognize the risks and opportunities created by restructuring. Regulators continued to focus on administrative oversight instead of facilitating competition, stalling the development of the retail market and impeding contracting by the utilities. The utilities failed to protect themselves from the risks they faced in the wholesale market. Consumers did not aggressively seek out new options for their electricity needs, preferring to remain passive consumers of electricity. Because these central actors continued to operate as if the stable and secure rules of regulation still held, they were woefully unprepared for the original price spikes in 2000, greatly exacerbating the extent of the crisis.

Much has already been done to address the most glaring causes of the California crisis. The long-term contracts signed by California, although expensive, have bought a measure of stability by reducing exposure to the spot market. New generation and conservation programs have eased the tight supply situation California faced, and market mitigation measures, including regional price caps imposed by FERC, have helped to reduce the threat of market power. Continued vigilance is needed, however. The exact causes of the crisis remain controversial, and because multiple, intertwined factors were simultaneously at work, it remains uncertain whether all the root causes of the crisis have been addressed. To ensure no repeat of the winter of 2001, policymakers will need to exercise caution as they contend with lingering uncertainties.

The most significant challenge facing policymakers is the need to develop a long-range plan for the California electricity sector. Because of the extent of the damage the crisis inflicted on market and policymaking organizations, California begins with nearly a clean slate. Policymakers can take the sector in a range of directions. They can increase the role of public ownership, or they can return to a more orderly world with regulated, vertically integrated monopolies. Alternatively, they can continue down the restructuring path that was interrupted by the crisis. Hybrid options include rebuilding the competitive wholesale market but with continued regulation of retail sales or only limited extension of retail competition to large industrial and commercial users.

The main tradeoff posed by these options is between the greater reliability and stability of regulated markets versus the efficiency gains and potentially lower costs made possible by competition. After enduring the crisis, the stability of regulation seems attractive. Nevertheless, competition has worked in other states and countries, and the advance of ever more efficient generation technology makes increasingly market-oriented policies inevitable at least in the generation sector. Effective deregulation, however, requires that California coordinate with federal regulators to develop a set of effective policies to prevent the exercise of market power.

As policymakers develop and implement a long-range vision, they should focus on a number of specific concerns:

- **Forge an early consensus.** Ambiguity and conflict concerning the future direction of California's electricity sector lead to market uncertainty. California risks repeating history if continued uncertainty stifles investment in critical infrastructure, leading to future shortages. Agreement on the broad outlines of a regulatory and market structure, even without the details specified, would do much to improve the investment environment.
- **Avoid the allure of quick gains.** The real benefits from competition do not accrue rapidly. It takes many years to improve the mix of operating plants, improve their operation, and develop more intelligent consumption patterns. In the early

years of the California restructuring experiment, actors focused excessively on reaping the gains of low wholesale rates. As a consequence, they failed to build the foundations of successful competitive markets, including retail competition, and left California exposed to the grave risks of the spot market. Pursuing quick fixes to California's current electricity woes risks again diverting attention from important fundamental reforms.

- **Improve demand management.** Expanding and institutionalizing demand-management programs is critical to making California's electricity sector more robust. Helping consumers make more intelligent consumption decisions can lower energy bills, improve efficiency, and help the environment. In addition, more active consumers of electricity facilitate a future transition to competition by increasing the benefits they can achieve from a wider selection of electricity offerings.

- **Reorganize the administration of energy policy.** Developing a post-crisis electricity sector will require a coordinated, comprehensive, and effective set of policies. The current set of electricity-related agencies, with their overlapping, conflicting, and ambiguous roles, are not up to this task.

Appendix

The Effect of Price Caps on Firms Exercising Market Power

As shown in Figure A.1, the common textbook example of rent control illustrates the adverse effects of a regulatory-imposed cap on prices, P_{cap}, in a perfectly competitive market. If left unregulated, this market will reach an equilibrium in which P^* is the market clearing price and Q^* units are sold. If P_{cap} is set below the market clearing price, P^*, suppliers reduce the amount they are willing to offer to the market from Q^* to $Q_{supplied}$ and consumers react to the lower price by increasing their demand to $Q_{demanded}$. The cap, thus, creates a shortage in the market equal to $Q_{demanded} - Q_{supplied}$. Assuming that this model of a competitive market correctly describes the California electricity market, the implication is that imposing caps would lead to blackouts, as generators would not be willing to supply all the electricity demanded at the capped prices.

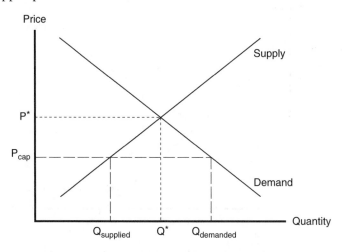

Figure A.1—Price Caps in a Competitive Market

A critical point largely missing from the debate is that *when firms are exercising market power*, this simple model does not apply, and price caps do not have this undesirable effect. As shown in Figure A.2, when a firm exercises market power, its marginal revenue—the increase in its revenues from selling one more unit of its product (e.g., a kilowatt hour of electricity)—is less than the current market price. This occurs because to attract one more buyer the firm has to lower its price. By making the additional sale, the firm earns the new (lower) market price for that unit, but at the same time, it is forced to accept lower revenues from all the units that it was already selling at the higher price. Consequently, the firm's marginal revenue curve is below the demand curve it faces. To maximize profits the firm will continue lowering (increasing) price and increasing (decreasing) sales until its cost for the last unit sold (e.g., its marginal cost or the supply curve) equals the marginal revenue earned for that last unit. Thus, a firm exercising market power will offer $Q_{market power}$ units on the market and charge a price of $P_{market power}$. If the firm did not exercise market power, it would offer Q^* units and charge a price of P^*. Consequently, a firm with market power sells fewer units and charges a higher price than a firm operating in a perfectly competitive market.

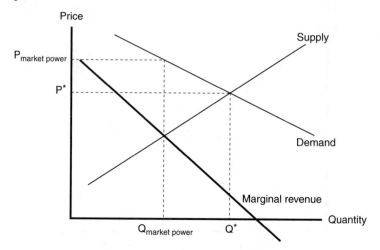

Figure A.2—Effect of Market Power

When a firm is exercising market power, a price cap, P_{cap}, changes the shape of its marginal revenue curve. Because it is no longer able to increase the price above the level of the cap, the cap determines the additional revenue it can earn from selling added units. As shown in Figure A.3 the cap flattens out the marginal revenue curve at the level of the cap. At the point that the cap intersects the demand curve, the marginal revenue drops down to the original level because the firm is again required to reduce the price of all units already sold to sell additional units. The firm then faces the new marginal revenue curve represented by the bold line. To maximize profits, the firm will sell all units for which the marginal revenue it earns exceeds its costs. Thus, it will produce Q_{cap} units, increasing output and lowering prices compared to the market outcome when it exercises market power.

Price caps remain a blunt policy instrument. Regulators can replicate the benefits of a perfectly competitive market only if the cap is set exactly to P^*, the equilibrium price under competition. The optimal, competitive price, however, is constantly changing with shifts in demand and supply conditions. Regulators have neither the information concerning market conditions nor the administrative capacity necessary to track these shifts. Thus, under typical circumstances, policies that promote vigorous

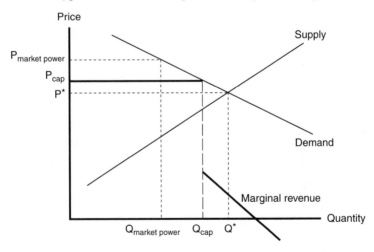

Figure A.3—Effect of Price Cap on Producers Exercising Market Power

competition in electricity markets have strong advantages over price caps in the long run. In contrast, price caps have advantages as a short-run policy when electricity supplies are tight enabling firms to exercise market power.

Even imperfect price caps can improve market outcomes and are not likely to exacerbate the problem. If the cap is set above the current market price, $P_{market\ power}$, then it has no effect, positive or negative, on market outcomes. As seen in Figure A.4, even price caps set below the optimal price, P^*, can improve market outcomes. The firm will produce Q_{cap} units, the point at which its costs equal the amount of the price cap. Although this result is not as efficient as the competitive market outcome, price caps increase the supply and lower the price of the good compared to the situation where market power goes unchecked. The one situation in which price caps do cause harm is if they are set below the current marginal costs of producers exercising market power ($MC_{market\ power}$ in Figure A.4). In this case, the price cap would reduce market supply and exacerbate shortages. In sum, when firms are exercising market power, regulators can improve market outcomes by setting any cap level between $MC_{market\ power}$ and $P_{market\ power}$. Although setting such a cap remains a difficult regulatory task, it is significantly less onerous than determining the optimal cap level.

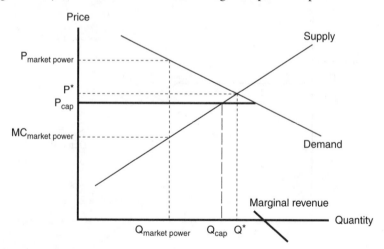

Figure A.4—Effect of Price Cap Below Competitive Level on Producers Exercising Market Power

References

Bay Area Economic Forum, *The Bay Area—A Knowledge Economy Needs Power,* San Francisco, California, 2001a, pp. 1–74.

Bay Area Economic Forum, *California at a Crossroads: Options for the Long-Term Reform of the Power Sector,* San Francisco, California, 2001b, pp. 1–56.

Bay Area Economic Forum, *The Economics of Electric System Municipalization,* San Francisco, California, October 2001c, available at http://www.bayeconfor.org/pdf/ElectricMuni.pdf.

Borenstein, S., *The Trouble with Electricity Markets (and Some Solutions),* Program on Workable Energy Regulation, University of California, Berkeley, 2001.

Borenstein, S., J. B. Bushnell, et al., *Diagnosing Market Power in California's Deregulated Wholesale Electricity Market,* University of California Energy Institute, Berkeley, 2001.

Brown, R. E., and J. G. Koomey, *Electricity Use in California: Past Trends and Present Usage Patterns,* Lawrence Berkeley National Laboratory, Berkeley, California, 2002.

Bushnell, J. B., and E. T. Mansur, *The Impact of Retail Rate Deregulation on Electricity Consumption in San Diego,* University of California Energy Institute, Berkeley, 2001.

Bushnell, J. B., and C. Saravia, *An Empirical Assessment of the Competitiveness of the New England Electricity Market,* University of California Energy Institute, Berkeley, 2002.

California Energy Commission, *Previous Energy Commission Peak Electricity Demand Forecasts,* California Energy Commission, 2001.

California Energy Commission, *The Summer 2001 Conservation Report,* California Energy Commission, 2002a.

California Energy Commission, *Power Plant Projects Recently Approved,* 2002b.

California Legislative Analyst's Office, *The 2002–03 Budget Bill: Perspectives and Issues—Reorganizing California's Energy-Related Activities,* Sacramento, California, 2002, available at http://www.lao.ca.gov/analysis_2002/2002_pandi/pi_part_5b_energy_an201.html.

California State Auditor, *Energy Deregulation: The Benefits of Competition Were Undermined by Structural Flaws in the Market, Unsuccessful Oversight, and Uncontrollable Competitive Forces,* Sacramento, California, 2001a.

California State Auditor, *Energy Deregulation: The State's Energy Balance Remains Uncertain but Could Improve with Changes to Its Energy Programs and Generation and Transmission Siting,* Sacramento, California, 2001b.

Cambridge Energy Research Associates, *Beyond California's Power Crisis: Impact, Solutions, and Lessons,* Cambridge, Massachusetts, 2001a.

Cambridge Energy Research Associates, *Short Circuit: Will the California Energy Crisis Derail the State's Economy?* UCLA Anderson Business Forecast, Los Angeles, California, 2001b.

Cicchetti, C., personal communication, University of Southern California, 2001.

Edison International Corporation, *2000 Annual Report,* Rosemead, California, 2001.

Federal Energy Regulatory Agency, *Market Order Proposing Remedies for California Wholesale Electric Sales,* Washington, D.C., 2000, pp. 1–77.

Federal Energy Regulatory Commission, *Report on Plant Outages in the State of California,* Office of the General Counsel, Market Oversight and Enforcement and the Office of Markets, Tariffs and Rates, Division of Energy Markets, Washington, D.C., 2001.

Friedman, L. S., and C. Weare, "The Two-Part Tariff, Practically Speaking," *Utilities Policy,* January 1993, pp. 62–80.

General Accounting Office, *Energy Markets: Results of Studies Assessing High Electricity Prices in California,* Washington, D.C., 2001.

Harvey, H., B. Paulos, et al., *California and the Energy Crisis: Diagnosis and Cure,* Energy Foundation, San Francisco, California, 2001, pp. 1–12.

Harvey, S. M., and W. W. Hogan, *On the Exercise of Market Power Through Strategic Withholding in California,* Center for Business and Government, JFK School of Government, Harvard, Cambridge, Massachusetts, 2001, pp. 1–76.

Hogan, W. H., *Electricity Market Power Mitigation,* Center for Business and Government, JFK School of Government, Harvard, Cambridge, Massachusetts, 2001a, pp. 1–23.

Hogan, W. H., *Electricity Market Restructuring: Reforms of Reforms,* Center for Business and Government, JFK School of Government, Harvard, Cambridge, Massachusetts, 2001b, pp. 1–29.

Joskow, P., "Deregulation and Regulatory Reform in the U.S. Electric Power Sector," in S. Peltzman and C. Winston, eds., *Deregulation of Network Industries: What's Next?* AEUI-Brookings Joint Center for Regulatory Studies, Washington, D.C., 2000, pp. 1–201.

Joskow, P., *California's Electricity Crisis,* Massachusetts Institute of Technology, Cambridge, Massachusetts, 2001.

Joskow, P. L., and E. Kahn, *Identifying the Exercise of Market Power: Refining the Estimates,* Cambridge, Massachusetts, 2001a.

Joskow, P. L. and E. Kahn, *A Quantitative Analysis of Pricing Behavior in California's Wholesale Electricity Market During the Summer of 2000,* National Bureau of Economic Research, Cambridge, Massachusetts, 2001b.

Kiesling, L., *Getting Electricity Deregulation Right: How Other States and Nations Have Avoided California's Mistakes,* Reason Foundation, Los Angeles, California, 2001.

Kumbhakar, S. C., and L. Hjalmarsson, "Relative Performance of Public and Private Ownership under Yardstick Competition: Electricity Retail Distribution," *European Economic Review,* Vol. 42, 1998, pp. 97–122.

Kwoka, J. E., *Power Structure: Ownership, Integration, and Competition in the U.S. Electricity Industry,* Kluwer Academic Publishers, Boston, Massachusetts, 1996.

McNamara, W., "FERC Enacts Major Constraints on Wholesale Market," *Scientech*, 2001a.

McNamara, W., "Five Reasons Why U.S. Deregulation May Be Stagnating (And Enron Isn't One of Them)," *Scientech*, 2001b.

Mowris, R., *California Energy Efficiency Policy and Program Priorities*, California Board for Energy Efficiency, Rosemead, California, 1998.

Oak Ridge National Laboratory, *Questions and Answers about Electric and Magnetic Fields Associated with the Use of Electric Power*, National Institute of Environmental Health Sciences, 1998.

Peltzman, S., and C. Winston, eds., *Deregulation of Network Industries: What's Next?* AEUI-Brookings Joint Center for Regulatory Studies, Washington, D.C., 2000.

Ross, S. A., *An Energy Crisis from the Past*, Institute for Governmental Studies, University of California, Berkeley, 1974.

Smith, D., "California Audit Finds Power-Plant Approvals by Energy Panel Often Delayed," *Sacramento Bee*, Sacramento, California, 2001.

Tucker, D. M., "California Officials Hope to Keep Averting Power Crisis," *The Business Press*, Ontario, California, 2002.

USC School of Policy Planning and Development, *Electricity Deregulation, Neo-Regulation, and Re-Regulation: What's Next?* University of Southern California, Los Angeles, 1999.

Vogel, N., "The California Energy Crisis; Cal-ISO Predicts 34 Days of Blackouts; Power: The Summer Could Be Filled with Rolling Outages If Consumers Don't Cut Consumption from Last Year's Levels, Officials Warn," *Los Angeles Times*, 2001, p. 26.

Wolak, F. A., *Designing a Competitive Wholesale Electricity Market That Benefits Consumers*, Stanford University, Palo Alto, California, 2001.

Wolfram, C., "Measuring Duopoly Power in the British Electricity Spot Market," *American Economic Review*, Vol. 89, 1999, p. 805.

About the Author

CHRISTOPHER WEARE

Christopher Weare is a research fellow at the Public Policy Institute of California. He has worked on the regulation of telecommunications, financial markets, and the electricity sector, focusing on the tradeoffs between efficiency and other goals. He is also researching the effects of information and communication technologies on local governance and citizen political participation. Before coming to PPIC, he was an assistant professor at the Annenberg School for Communication at the University of Southern California and a visiting assistant professor at the University of California, Berkeley. He also spent one year as a Congressional Fellow in the U.S. House of Representatives. He holds a B.A. in government from Harvard College and a Ph.D. in public policy from the University of California, Berkeley.

Related PPIC Publications

California's Infrastructure Policy for the 21st Century: Issues and Opportunities
David E. Dowall

Building California's Future: Current Conditions in Infrastructure Planning, Budgeting, and Financing
Michael Neuman and Jan Whittington

PPIC publications may be ordered by phone or from our website
(800) 232-5343 [mainland U.S.]
(415) 291-4400 [Canada, Hawaii, overseas]
www.ppic.org